New Deal Medicine

Dear Tom,

It was the highlight of my brief Dallas sojourn to meet you. Perhaps we'll meet on the red road Again soon! —

Gabriel

New Deal Medicine

The Rural Health Programs
of the Farm Security Administration

MICHAEL R. GREY

THE JOHNS HOPKINS UNIVERSITY PRESS
Baltimore and London

Johns Hopkins Paperbacks edition, 2002
9 8 7 6 5 4 3 2 1

The Johns Hopkins University Press
2715 North Charles Street
Baltimore, Maryland 21218-4363
www.press.jhu.edu

Library of Congress Cataloging-in-Publication Data will be found
at the end of this book.
A catalog record for this book is available from the British Library.

ISBN 0-8018-6917-X (pbk.)

The Farm Security Administration documentary photographs that appear throughout
the book are located at the Prints and Photographs Division, Library of Congress,
Washington, D.C.

For my parents

The program of the FSA is so little understood by the general public and condemned by some, that I wish for just a few moments to discuss it.
CARL VOHS, M.D., PRESIDENT, MISSOURI STATE MEDICAL SOCIETY, 1944

You have to know what the whole system's about to make a good judgment, and that includes knowing what went on in the FSA days. And if people don't believe that, then they better forget about democracy.
THOMAS GARLAND MOORE JR., FARM SECURITY ADMINISTRATION, 1979

Contents

Preface

In this book, I reconstruct a largely neglected and highly creative experiment in twentieth-century federal health care policy. In doing so, I hope to illustrate some of the enduring themes that punctuate the debate over the proper role of government in health care, the response of the medical community to health reform, and the successes and failures of the strategies used by the New Deal's Farm Security Administration (FSA) as it attempted to achieve its goals.

The agency's creation of a voluntary, public-private program established important precedents for future federal health policy initiatives on behalf of low-income citizens. Features of the FSA programs bear a striking similarity to recent developments in the public and private sector. For example, FSA policies prefigured later changes in health care delivery in the United States, including capitated financing, the central role of the general practitioners (and nurses) in managing illness and controlling access to specialized services, negotiated agreements with physician groups, consumer participation, and peer and utilization review.

The FSA medical care programs do not feature prominently in the historiography of twentieth-century American agriculture or medicine. With the exception of my own work, very little has been written about these programs in the fifty years since they ended. Sidney Baldwin's *Poverty and Politics,* which was published some thirty years ago, remains the most detailed history of the FSA as a whole. However, Baldwin summarizes the FSA's medical care programs in a scant two paragraphs, although their important role in the agency's overall rehabilitation effort is given appropriate credit. Two other sources cover the FSA programs in more depth. Frederick Dodge Mott and Milton Roemer dedicated a chapter in their 1948 book, *Rural Health and Medical Care,* to the FSA medical programs. This book was written on the heels of the program's demise, and both Mott and Roemer are central characters in the history of the FSA medical programs. In my view, this monograph represents an essential primary source, rather than a true secondary analysis of the FSA programs. The medical care programs also feature prominently in John Stoeckle and George Abbott White's *Plain Pictures of Plain Doctoring*. However, Stoeckle and White's primary intent was to use the FSA's documentary photographs to provide a visual portrait of general practice during the Great Depression. Much of their information on the FSA was based on my undergraduate research on the agency's health care programs.

Because the secondary literature on the FSA medical care program is so sparse, nearly all of the material on which this book is based comes from the agency's archival records, from medical journals of the day, and from the recollections of a remarkable cadre of individuals who participated in the FSA medical care programs. The FSA was a program with national reach that lasted over a decade. Consequently, the development, structure, impact, and response to the medical care programs varied (sometimes considerably) from region to region, from one locale to another within a state, and over time. This complicated my task of presenting the history of the FSA medical care program and required hard choices in selecting what to include in the narrative. My decisions were guided by the desire to provide a balanced and comprehensive view of the medical care programs without sacrificing the subtleties that emerge only when one looks at the program as it operated at the local level. These choices were also driven by existing resources. The evidentiary trail of a program that existed half a century ago is sometimes a wide and firm road and sometimes a barely visible footpath. To borrow a term coined by sociologist Richard Couto, what emerges is perhaps best characterized as "political archaeology" rather than a simple historical narrative.[1]

The personal reminiscences quoted in this book were chosen to provide a more personal view of key points in the narrative. It is also my hope that they give the reader the flavor of a period in our history that is rapidly passing beyond living memory. Most of those interviewed were deeply involved in the New Deal; a number were U.S. Public Health Service medical officers, and all were direct participants in the FSA medical care programs. Nearly all of the physicians interviewed would view themselves as progressives within the generally conservative house of medicine. They are not intended to be representative of all doctors involved in the program. However, like many who were drawn into the New Deal, they shared a vision of a better society and a belief that the federal government had a responsibility to ensure this vision for all Americans. Their remembrances of the FSA medical care program, blemishes and all, enhance the view presented by the archival record.

The FSA provided a widely dispersed group of rural citizens and their doctors direct experience with health insurance as a means of improving access to medical care and limiting the financial burden of illness. My view that the FSA programs were both an effort to solve pressing problems and an attempt to make permanent changes in the way in which health care was delivered places the agency in the center of today's national health policy debate.

Acknowledgments

While the Farm Security Administration medical care programs have held my attention for most of my adult life, the time I spent researching and writing about the FSA has been punctuated by long stretches dedicated to practicing medicine. It is time now to thank those institutions and individuals who have made this book possible.

In an age of virtual communication of all stripes, there is still satisfaction in delving into archives filled with yellowed and dusty manuscripts, letters, and various sorts of materials. I owe a great debt to the following archives and libraries, and to the staff that assisted me during my visits to them. Most of the primary government documents are located in the National Archives in Washington, D.C., as well as in the Federal Records Centers of the National Archives and Records Administration. The three regional Federal Records Centers in Woburn, Massachusetts; Berkeley, California; and Seattle, Washington were instrumental in expanding my research on the FSA in order to achieve geographical diversity and depth on the FSA health programs. Additional primary and secondary materials were found in the Bancroft Library of the University of California at Berkeley, Yale University's Sterling Memorial Library, and the extensive library system at Harvard University.

The medical journals I used to present a running commentary on the FSA from the perspective of ordinary physicians as well as the leadership of organized medicine were found in the medical libraries of the University of Washington, Yale University, and the University of Connecticut Health Center. Dr. Ralph Arcari, director of the Lyman Maynard Stowe Library at the University of Connecticut Health Center, has never stinted in offering me the assistance of either himself or his staff when I needed it.

The secondary sources used to broaden my understanding of the history of

American agriculture were found in the library systems of two land-grant institutions: the University of Connecticut and the University of Washington. The Canadian National Archives in Ottawa gave me free access to the personal and professional papers of Frederick Dodge Mott.

The outstanding collection of documentary photographs of the FSA are located at the Prints and Photographs Division of the Library of Congress in Washington, D.C. Selected FSA prints are located at the Fogg Art Museum at Harvard University.

The personal reminiscences of the physicians, nurses, and FSA supervisors afforded a unique perspective on the Great Depression, the New Deal, and the achievements of the FSA. They also helped me appreciate the FSA's ties to developments in health care in the United States and abroad in the years following the program's termination. The following individuals gave willingly of their time and many also lent me their personal papers: Lorin Kerr, Leslie Falk, Milton Roemer, John Newdorp, George Silver, Allen Koplin, Pauline Koplin, Henry C. Daniels, Harold Mayers, and Helen Johnston. I owe a special note of gratitude to the irrepressible Thomas Garland (T. G.) Moore Jr., whose vivid recollections of the Great Depression and whose lifelong commitment to social justice have stayed with me ever since we first met. Several of the epigraphs that appear before each chapter are based on talks I had with him. Before his death in 1980, Dr. Frederick Dodge Mott graciously lent me his personal records. They now reside in the Canadian National Archives, a testimony to Dr. Mott's contributions to public health and medical care in that country. I was told I was the first researcher to request access to a collection of which our neighbors to the north are quite proud. I hope someday that the remarkable career of Dr. Mott will draw the interest that it deserves. His family, especially his son, Andrew Mott, clarified for me many details of his later career.

My research was greatly aided by the generosity of several foundations over the years. As a senior medical student, through a Smith Kline Beckman Creative Perspectives in Medicine grant I was able to interview a number of participants in the FSA medical care programs. As a Robert Wood Johnson Clinical Scholar at the University of Washington, I interviewed more participants and widened my research to include the agency's migrant health programs, resettlement projects, and experimental health plans. Early manuscript preparation and corroborative research were supported by a grant from the National Endowment for the Humanities. I wish to thank in particular Daniel Jones of the Health Sciences and Technology Division of Research Programs at the National Endowment for the Humanities for his encouragement and enthusiasm for this project. The final stages of manuscript preparation would have been indefinitely delayed if it had not been possible to at least partially limit my clinical and academic responsibilities. For this, thanks go once more to the

Robert Wood Johnson Foundation for awarding me a Generalist Physician Faculty Scholars grant. Twice in my career I have benefited from this foundation's affirmation of the value of the humanities in medical education and health policy, and in its potential to make medicine a more humane endeavor. The foundation clearly believes, as I do, that the gulf between medical practitioners and historians of medicine is a loss for both communities.

Over the years, I have published several articles relating to the FSA medical care programs. The following journals have graciously agreed to let me incorporate selected excerpts in this book: the *Annals of Internal Medicine,* the *Journal of the History of Medicine and Allied Sciences,* and the *Journal of the American Public Health Association.*

I owe my greatest debt to those colleagues and friends who have given freely of their time, expertise, and unflagging support throughout my career. Dr. Mark Smith first brought to my attention the medical photographs of the FSA's Historical Division while we were undergraduates at Harvard. It was then that I realized that the literature on twentieth-century American medicine was largely silent on this remarkable experiment in federal health care delivery, and I undertook to rectify that situation. My earliest research on the FSA was nurtured by the enduring enthusiasm shown for the idea by my thesis advisor and mentor, Dr. John Stoeckle; by my teacher and friend, George Abbott White; and by Professor Barbara Rosenkrantz. Barbara's unwavering commitment to the highest standards of historical scholarship has at various points prodded, shamed, and energized me to write a book that I hope is worthy of the story it tells. Drafts of this manuscript have also been read by Professors Elizabeth Fee and David Rosner. Their gentle criticism and their knowledge of twentieth-century American medicine and health policy made this a much better book than would otherwise have been the case. Where many aspiring physician-historians would be without the thoughtful advice of John Harley Warner I cannot guess. I, too, have benefited greatly from his thoughtful mentoring and balanced approach to medical history.

Colleagues and friends I met while a Robert Wood Johnson clinical scholar in Seattle helped me see how my historical research could illuminate more contemporary health policy issues. They also urged me to consider the audience I hoped to reach in my work. My program director, Tom Inui, is among the most important influences in my career, and I am very grateful for his continued support of my desire to merge historical scholarship with the demands of being a physician. My participation in two graduate seminars at the University of Washington was critical in sharpening my analysis of both the New Deal and the FSA. Particular thanks are due to Professors Keith Benson, George Behlmer, and Robert Burke—and their graduate students—for helping me to think more like a historian and write less like a physician. Over the years my involvement with the Robert Wood Johnson Foundation has introduced me to a cadre of like-minded physician-humanists whose example has been of professional and intellectual benefit to me. Without the example and guidance of Kenneth Ludmerer, Joel Howell, Barron Lerner, Howard Markel, and Robbie Aronowitz, my own work would have been much harder to carry on. Robbie's career and mine have been in parallel for some time now; he has been an inspiration and a good friend.

I could not have completed this project were it not for the support of many of my colleagues at the University of Connecticut School of Medicine. A special note of appreciation is in order for the following people: Robert U. Massey, Richard M. Ratzan, Eileen Storey, Henry Schneiderman, Tony Voytovich, David Courtwright, and Jay Healey. The memory of the late Jay Healey—with whom I shared a love of teaching, William Carlos Williams, Walker Percy, rock 'n' roll, and baseball (although not necessarily in that order)—continues to guide my academic career. In an era when academic health centers have been hard pressed to preserve their educational, research, and clinical missions, I have been fortunate that a succession of deans, department chairs, and division chiefs at the University of Connecticut School of Medicine have shared my belief that medical humanities deserves a place in the education of our students and residents.

Finally, although they are too numerous to mention by name, I would like to add a note of general appreciation to my colleagues in the Department of Medicine, the Division of General Medicine, the Division of Occupational/ Environmental Medicine, and the Department of Community Medicine and Health Care. Finally, Camille Prentice and Sandy Daversa in the Primary Care Internal Medicine Program were always willing to help out. The final revision of the manuscript was undertaken while I was on sabbatical at the University of Newcastle in Australia. The many kindnesses shown to me and my family by our Aussie friends during our six months in the antipodes, and the inspiration provided to me by the faculty of that remarkable educational institution have left me with many wonderful memories. The generosity of John Hamilton, Isobel Rolfe, and Dr. Tony Brown—and their colleagues in the Programme Evaluation Unit and in the discipline of occupational health— remains among the most indelible impressions I have.

In the past several years, I was assisted by three able research assistants: Tarik Kardestuncer, Glenn Konopaske, and Tom Carty III. I have been blessed to have worked in the final stages of the book with a remarkable editor, Kennie Lyman. Kennie's editorial legerdemain helped me craft a much tighter narrative, and her steady enthusiasm for the project in its late stages helped me overcome inertia. I undoubtedly tried the patience of Jackie Wehmueller at the Johns Hopkins University Press, but I am very thankful that the missteps of a first-time author never caused her to abandon her eagerness for this project.

I hope that this book affirms the confidence each of these individuals has shown in me over the years. These acknowledgments stated, any errors that remain are certainly my own.

My final comments are for my friends and family. The following people deserve acknowledgment for the friendship and encouragement they extended over many years: Mark Smith, Michael Tierney, Susan Willis, James Judge, Meredith Fields, Thomas Barber, Debra Williams, and Melissa Smith. I want to thank especially Robert Podoloff, David Bachman, Andrew Dodds, and Jefferson Singer for a lifetime of friendship and wisdom. My parents, Ross Maxwell Grey and Marion Perkins Grey, and my three sisters, Jennifer, Elizabeth, and Julie, have always been supportive. I suspect that my father's North Carolina roots may have played a role in my focusing on a program that had an especially significant impact in that region. Although my grandmother, Helen Wallace Perkins, disliked FDR and positively detested Eleanor, I have no doubt that her stories of the Great Depression awakened my interest in what is one of the defining periods in our nation's history.

Preparing a book for publication is a task that I seriously underestimated when I first began. My wife and my two boys have felt the chronic disruption of our lives that bringing this book to fruition made necessary. Nicholas and David cheerfully interrupted me whenever I neglected to lock the study door, and their visits reminded me that my most important job in life was not as a historian or as a physician. I can only imagine how much my wife, Eda diBiccari, has endured over the past several years. I have undoubtedly pushed the limits of her patience, but never the limits of her love.

New Deal Medicine

Introduction

An Experiment in Democracy and an Entering Wedge

There's no doubt that in many instances it made the difference between life and death. Today we would call it Mickey Mouse, you know it was put together with baling wire. But it was what there was. Otherwise people were scrounging and did whatever they could on whatever little money they had, which means it helped a desperate situation.
THOMAS GARLAND MOORE JR., 1979

Thomas Garland Moore was working his mules on a dusty stretch of road in West Texas when an official-looking man came hunting for "a feller named Tom Moore." The year was 1933 and America was in the grip of its most terrible economic crisis. Like many of his fellow citizens, Moore had spent his share of time on the road in search of work. A teacher by vocation, he wandered around the parched southwestern prairie in the bleakest years of the Great Depression, finding temporary employment as an oil company bookkeeper, watermelon packer, and cotton gin hand. Moore paused from his road maintenance work as the man approached and thought to himself:

> "Well I'm fired." I got out on top and dusted the white rock off me. He says, "We see from your papers that you are a bookkeeper and a teacher . . .We got a job down here in the clerical capacity and wonder if you'd be interested?" "How much you pay?" I asked and he said, "Twelve bucks."And I said, "You hired yourself a man." I hung up my lines and went. And that's the way I got started in the government.[1]

For Moore this was the start of a long association with the New Deal, one he began as an accountant in West Texas for the Federal Emergency Relief Administration (FERA) and ended as a regional supervisor for a system of migrant labor camps funded by the federal government.

Moore's familiarity with the Great Depression and his personal stake in the success of Franklin Delano Roosevelt's New Deal were experiences he had in common with millions of Americans. The era of American history between the stock market crash in October 1929 and the end of World War II has been a rich source of inspiration and reflection for two generations of historians, sociologists, and artists. Writers such as James Agee and John Steinbeck and

Migrant mother. Nipomo, Calif., 1936. Dorothea Lange.

artists such as Diego Rivera were profoundly influenced by the period. Ironically, our collective image of the Great Depression has been formed by photographs taken by one of the New Deal's least-known social welfare agencies, the Farm Security Administration, or FSA.[2] Although the FSA has languished in relative historical anonymity, the creative talents and enthusiasm of the photographers, journalists, and scholars employed by its Information Division provide us with some of the most lasting images from this period. The haunting documentary photographs of Dorothea Lange, Walker Evans, and their colleagues give us our most poignant and vivid images of the Great Depression. So powerful are many of these photographs that when they were first seen, they galvanized public debate around poverty and social justice to a degree hardly matched in our nation's history.

AN EXPERIMENT IN DEMOCRACY:
THE FARM SECURITY ADMINISTRATION

The FSA was one of several New Deal agencies whose purpose was to provide for the "bottom third" of Americans, those left indigent by the massive economic collapse of the 1930s. Its goal was the rehabilitation of "1,700,000 farm families who in 1936 were trying to pay rent, operate their farms, and feed and clothe themselves on an average income of only $500 a year."[3] From 1935 to 1946, the FSA (and its short-lived predecessor, the Resettlement Administration, or RA) provided low-interest loans to impoverished farmers, share-croppers, and farm laborers to allow them to buy and farm their own land. It established a broad network of farm labor camps in regions dependent on migrant labor and supported a decentralized technical and educational staff in an innovative, if ultimately inadequate and transient, assault on the sources of rural poverty. The vigor with which the FSA pursued its mandate established its reputation as one of the most socially conscious of all New Deal programs. The agency was a magnet for highly motivated individuals who, like T. G. Moore, shared a passion for social change. Their experience working for the FSA nourished their idealism while it honed skills that allowed them to put their ideals to work.

The FSA elicited the bitter hostility of those opposed to the New Deal. For them, the agency was emblematic of the ill-conceived social engineering characteristic of the New Deal. The FSA's efforts to promote economic self-sufficiency among indigent rural families drew particular criticism from commercial farm interests and their allies in the agricultural establishment, including land grant colleges and the Department of Agriculture's own Agricultural Extension Service (AES). This alliance was sometimes referred to as the "cliental bloc" because its members were "clients" of conservative politicians who represented their interests and benefited from their voting power. However, since these politicians were known as the "farm bloc," it is clearer to refer to this alliance by the term that is used to describe it today—the farm bloc. In the most comprehensive institutional history of the FSA, historian Sidney Baldwin wrote that this coalition represented "an important source of danger to the life of the FSA."[4]

Critics of the FSA, both at the time and since, have contended that its results never came close to matching its rhetoric.[5] The job that it faced was overwhelming. Today it is difficult to imagine the scope of the economic catastrophe and the destitution and hopelessness of the Depression's victims. Judging the FSA only by how nearly it met its lofty goals—by the number of rehabilitation loans it made, the number of indigent families it resettled on more productive land, or the number of migrants it kept from starving in roadside squatter villages—risks failing to acknowledge the extent of its impact on rural America during the Great Depression.

In *Ain't Gonna Let Nobody Turn Me Around,* Richard Couto traces the links between federal bureaucracies that emerged in three distinct periods of unusually far-reaching federal involvement in American society: the Freedman's Bureau during the post–Civil War Reconstruction Era, the Farm Security Administration during the Great Depression, and the Office of Economic Opportunity during the Kennedy-Johnson administrations' War on Poverty. Couto describes these agencies as "heroic bureaucracies," which he defines as bureaucracies that "attribute more human dignity and worth to members of subordinate groups."[6] This characteristic, as well as their innovative efforts to redress chronic economic inequality, make these agencies targets for those intent on preserving the status quo. The history of the FSA conforms to this analysis.

The FSA could provide for only a fraction of those 1.7 million families it sought to assist. Despite nearly continuous opposition and budgetary constraints, however, the agency provided physical, economic, and psychological sustenance to some of the nation's poorest citizens during its twelve-year history. Baldwin and other agricultural historians believe that the FSA's leaders "improved the human condition for hundreds of thousands of destitute farm families, conceived and successfully applied unique social innovations, maintained their agency's viability, and explored the parameters of the possible."[7]

NEW DEAL MEDICINE:
THE MEDICAL CARE PROGRAMS OF THE FSA

In his second inaugural address in 1937, President Roosevelt spoke of one-third of the nation as "ill-housed, ill-clad, and ill-nourished."[8] The FSA learned in the course of its rehabilitation work that many poor rural families were also just plain ill. In an effort to respond to this fact and to protect its investment, the FSA made medical care a core element of its rehabilitation programs. The FSA medical care program is credited by several historians as being one of the agency's most lasting and creative achievements. In 1950, rural sociologist Olaf Larson wrote that of all the FSA's rehabilitation efforts, perhaps the greatest gains were in the physical health of FSA families. Twenty-five years later, historian Elizabeth Etheridge cited the FSA's nutritional programs and its massive adult education campaign as one of several factors leading to the eradication of pellagra as a significant public health problem in the South. [9]

Although humanitarian ideals were undoubtedly at work, publicly FSA leaders emphasized an economic rationale for the agency's involvement in medical care delivery, echoing a common historical theme used to promote government intervention in health and welfare matters.[10] According to the FSA, nearly half of all loan defaults were attributable to sickness. Health surveys done by the FSA—often with the assistance of one of the agency's most steadfast allies, the U.S. Public Health Service—indicated that chronic, debilitating physical conditions often hindered effective rehabilitation. The contribution of good health to the successful economic rehabilitation of low-income farmers was a persuasive argument used by the FSA to justify its health program to both its supporters and its detractors.

At its peak in 1942, the FSA had over 650,000 poor rural farmers enrolled in prepaid medical cooperative plans operating in more than a third of all rural counties in the United States. Working with the guidance of the FSA's Health Services Branch, FSA relief and rehabilitation supervisors assisted families, often referred to as "clients" or "borrowers," in working out collective arrangements with local medical practitioners for general medical services. More comprehensive care—in the form of pharmaceutical, surgical, or hospital services—was commonly but not invariably included in these plans. However, the FSA's focus was clearly on general physician care, coupled with an abiding emphasis on prevention and health education.

Low-income farmers were not the sole target of the FSA's rehabilitation effort. The farm labor shelter program provided housing and a sanitary environment to those migrant families fortunate enough to find refuge in the federal camps. In a development that paralleled the medical cooperative program for low-income rehabilitation borrowers, the FSA established a migrant health program that provided acute care, hospitalization, and preventive services to a migrant population swollen by the combined effects of national drought, farm mechanization, and the near collapse of the nation's farm economy. The FSA also experimented with more controversial health care delivery models. These included two statewide insurance plans in the drought-stricken states of North Dakota and South Dakota; the medical cooperatives the agency sponsored in its resettlement communities; and after 1942, a series of countywide experimental health plans whose comprehensive medical services were not restricted to low-income FSA clients.

The FSA first promoted its medical care program, in which the federal government was the fiscal intermediary, at a time when organized medicine opposed the very notion of third-party payment for medical care. The agency's success can be traced to characteristics of the plans themselves and to the cir-

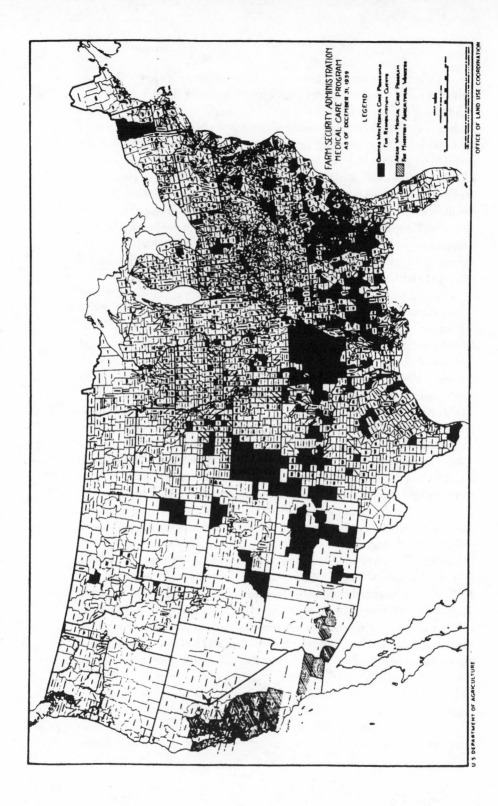

FARM SECURITY ADMINISTRATION
MEDICAL CARE PROGRAM
AS OF DECEMBER 31, 1939

LEGEND

Counties with Medical Care Program
for Rehabilitation Clients

Areas with Medical Care Program
for Migratory Agricultural Workers

U S DEPARTMENT OF AGRICULTURE

OFFICE OF LAND USE COORDINATION

cumstances surrounding their implementation. For example, the FSA targeted groups that had few resources to pay for medical care and provided timely
economic support to hard-pressed rural physicians. In addition, it adopted
many of the tenets of greatest concern to physicians, such as free choice of
physician and voluntary participation. Finally, each medical care plan was negotiated with local or state medical associations. This approach gave physicians substantial but not absolute control over the operation of the medical
care plan. For a time, many local physicians viewed the FSA programs as less
threatening than some of the alternatives then being considered, particularly
compulsory national health insurance. As FSA historian Sidney Baldwin
notes, cries of "socialization of medicine" against the program were virtually
absent during these early years."[11]

Many rural physicians had watched their neighbors struggling to hold onto
failing farms in the face of the overwhelming economic downturn. Humanitarian instincts forced them to consider the FSA program, whatever their ideological qualms, because it offered a way to provide health care to their medically needy neighbors. There were powerful economic incentives as well.
Until the dramatic post–World War II growth in physicians' income, the financial status of rural physicians was often precarious, and barter and bad
debt were the lot of many a country doctor. Matters were made significantly
worse in the early years of the Depression as collection rates plummeted below 50 percent of billings.[12] The FSA program was a financial shot in the arm
for rural general practitioners at a time when county, state, and philanthropic
resources were overwhelmed by the collapse of the national economy. For its
part, although the American Medical Association (AMA) was distrustful of
third-party intervention in health care and on record until 1938 as opposing
even voluntary health insurance, it deferred to the financial needs of its rural
guildsmen during the early years in which the FSA programs expanded.

The FSA health programs also promoted the social and professional status
of allopathic physicians. At a time when competing theories of medical practice, such as osteopathy and chiropractic, had strong appeal in rural America,
FSA policy stipulated that only licensed doctors of medicine were eligible for
participation in its medical care plans. The FSA programs also offered other
more tangible prerogatives, such as a guaranteed patient base and hospital
privileges. In exchange for fiscal and administrative concessions, allopathic
physicians gained legitimacy through their association with a large, and for a
time popular, federal program.

For the FSA leadership, genuine conviction as well as pragmatic politics
dictated an alliance with organized medicine. By the 1930s, physicians exer-

Facing page: *Counties with FSA medical care cooperatives, about 1939.* USDA,
Farm Security Administration, Annual Report of the Chief Medical Officer, *1939.*

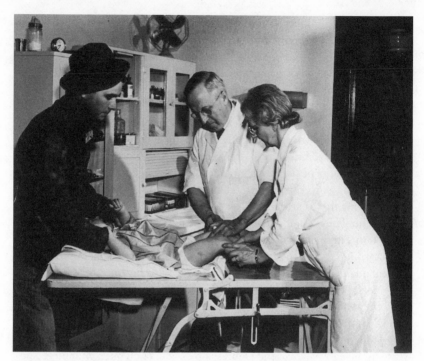

*Little girl who was bitten by a dog receiving antihydrophobia vaccine.
Chaffe, Mo., February 1942. John Vachon.*

cised substantial influence in defining what constituted good medical care.
FSA medical leaders, nearly all of whom were physicians and most of whom
were commissioned officers in the U.S. Public Health Service, shared these
basic assumptions. Indeed, all the players in the FSA story—the agency, its
clients, and organized medicine—shared a belief in the intrinsic value of medicine. They believed it could redress both individual and social ills and that
more medical care was by definition better. Whatever doubts we may have today about the efficacy of allopathic medicine, during the 1930s and 1940s its
value was unquestioned and faith in it underlay public, political, and professional discourse. "In the 1930s," writes historian Charles Rosenberg, "the fundamental problems of health care were not perceived as intrinsic to scientific medicine; instead they seemed to lie in maldistribution of the real benefits
that medicine could provide." [13] The divisive issues then (and now) were in
what direction and under whose control changes in the delivery of medical
care should proceed.

Although Farm Security Administration programs benefited from their
consonance with the financial, professional, and personal interests of rural

doctors, many medical societies and individual physicians were ill at ease with the expanding federal presence in medical care delivery. The FSA programs' broad public health orientation—one that included nurses, sanitarians, health educators, and home management supervisors, along with doctors in the provision of health services—was an additional cause for concern. Finally, the agency's experimentation with organizational, administrative, and fiscal approaches to health care delivery discomfited many organized medical groups, including the American Medical Association.

For several years, FSA leaders alleviated these anxieties by stressing the economic value of a medical care program in the overall picture of rural rehabilitation. Since most participating physicians viewed the FSA program as an emergency remedy to provide medical treatment to a well-defined rural indigent group, rather than a model for national health insurance, FSA leaders did little to suggest otherwise. This strategy allowed the agency to deflect for a time the concerns raised by those who feared that it was part of the New Deal's agenda to pass national health legislation.[14]

After 1942, the economic, professional, and humanitarian threads that once tied local physicians to the FSA began to unravel, and membership in the FSA medical care programs dropped sharply. The reasons behind these developments are central issues whose consideration will shed light on physicians' motivations during periods of health care reform and the obstacles facing the federal government in its efforts to promote a national health care system. To some extent, the program's decline reflected changing social, economic, and political realities that played out differently among farmers, physicians, and the federal government and that need to be understood in order to properly consider the impact and legacy of the FSA.

As the American economy prepared for war in the years prior to Pearl Harbor, the lingering effects of the Great Depression on a weary nation mercifully dissipated. This was especially true for physicians, who saw their incomes rise dramatically. Rural physicians, who were once willing to participate in a program that paid the medical bills of a group that had traditionally received little care and paid for even less, now had little financial incentive to continue their partnership with the FSA. This rising economic tide did not carry all boats, however. Many of the most severely needy rural families—especially women, children, and the elderly—were bypassed by the prosperity that the rest of the nation enjoyed. For these families, the advantages of a prepaid medical care plan subsidized by the federal government remained.

World War II's demographic impact on the country also played a role in the difficulties experienced by the FSA medical care programs after 1942. Physicians and nurses were drafted in large numbers, and many farmers abandoned their communities for the armed forces or urban war industries. With fewer physicians and a decreasing number of rural families, it would seem that the

FSA health programs died a natural death. At the time, in fact, FSA leaders focused much of the blame for the programs' decline on the war. Two years after the termination of the Farm Security Administration, a book written by former FSA Chief Medical Officer Fred Mott and his assistant, Milton Roemer, placed much of the onus for the agency's problems on the war. Mott and Roemer's thesis has been echoed by those few historians who have commented on the FSA medical care programs.[15] However, this is only part of the story.

The changing demographic and economic environment coincided with political changes in Washington. Growing congressional opposition to the New Deal's legislative activism, organized opposition to the FSA by the farm bloc, and Roosevelt's preoccupation with the war added to the agency's woes. Acrimonious appropriations hearings and a series of ill-willed congressional investigations into the FSA further sapped the energies of the agency's leaders and forced them to scale back its activities beginning in 1942. Under the guise of wartime restructuring, for example, the FSA was stripped of its migrant programs in 1943, although the migrant health programs continued to be supervised by FSA Health Services personnel. As the FSA lost ground in Washington, many of the old guard that had guided it during the so-called golden years between 1937 and 1942 left, or were driven out to pacify agency critics.[16]

During the war, FSA leaders tried to adapt the agency's programs, including its health care initiatives, to shifting national priorities. The agency assisted the Red Cross in organizing first-aid classes and conducted physical examinations of youths enrolled in the Civilian Conservation Corps and National Youth Corps. The FSA's experience in transporting and sheltering farm workers also made it a natural choice to coordinate the transportation and sheltering of agricultural laborers who were brought to the United States as replacements for American farm labor lost to the war effort. As part of this emergency farm labor shelter program, the FSA participated in one of the most shameful episodes in our nation's history: the forced removal of Japanese Americans from the Pacific Coast. In the panic that followed the bombing of Pearl Harbor, nearly 110,000 naturalized citizens and their American-born children were evicted from their communities, boarded onto trains, and placed in internment camps in the nation's interior. For FSA personnel, who viewed the agency as a champion of the underdog, this was an especially bitter duty. [17]

The FSA stumbled through the war, cobbling its rehabilitation programs to fit new priorities mandated by an increasingly unfriendly Congress. In 1946, Congress finally terminated the FSA, bringing to an end the agency's experiment in rural health care delivery. The FSA's remaining programs were transferred to the Farmers Home Administration, a newly created agency whose legislative mandate was far more circumscribed and whose involvement in rural health issues would not be meaningful until the mid-1970s. This was the end of the FSA health care programs.

Neither the demographic changes and economic prosperity of the war years nor the hostility of Congress fully explain the demise of the FSA experiment in health care delivery. At the heart of the program's decline was the tension between the FSA's desire to expand the role of the federal government in the provision of and payment for medical services, and the medical profession's fear that this would undermine the sanctity of the doctor-patient relationship and the profession's authority over the practice of medicine. Many of the agency's problems in fact can be traced to its earliest days. Closer examination of the difficulties the FSA experienced during its period of rapid growth reveals that individual physicians and county and state medical associations frustrated the agency all along. For example, while the total number of medical care units increased up to 1942, many plans ran into trouble before the nation's entry into the war, and a substantial number failed to materialize at all because of physicians' opposition.

On a national level, the AMA had always been wary of outside involvement in medical care delivery, especially from the federal government. These views were echoed to a greater or lesser extent at the state and local level. The crisis caused by the Depression for both patients and doctors, however, forced physicians and the groups that represented them to consider measures that they would have flatly rejected in less extraordinary circumstances. Although AMA leaders were unwilling to embrace the FSA proposals, the dire social and economic conditions compelled them to pass along the ultimate decision-making power to state and local medical societies, which varied widely from the start in their responsiveness to the program.

As private health insurance gained a foothold on the American medical landscape in the later part of the 1930s, the AMA was forced to reconsider its position. Rather than fight against any medical care program that diverged from traditional fee-for-service medicine—a battle it had already lost—organized medicine focused its energies on promoting physician-sponsored or commercial insurance as alternatives to government-sponsored plans. Once this occurred, many of those rank-and-file physicians who had been willing to cooperate with the FSA at the local and state level realigned themselves to support the position taken by their organized national voice.

VOLUNTARY HEALTH INSURANCE AND THE CHANGING STATE OF MEDICAL CARE IN THE UNITED STATES

The FSA was not the only federal bureaucracy involved in medical care and health-related activities in the era bracketed by the Great Depression and the end of World War II, although the magnitude of its commitment surpassed that of other New Deal agencies.[18] This was a period of broad social experimentation in both the private and public arenas, and a number of innovative health care delivery plans date from this time. Several private philanthropies gave both attention and money to improving health care for the needy, some of them rural communities. Notable among these efforts were those of the Duke Endowment in the South, the Kellogg Foundation in Michigan, the Commonwealth Fund, and the Julius Rosenwald Fund, which had a particular interest in the health and medical care of African Americans. For the most part, however, philanthropic involvement in health care delivery did not begin in earnest until the 1940s and was concentrated only in selected areas of the country. It also gave greater emphasis to hospital construction and public health programs, rather than medical care delivery.[19]

The majority of these health care programs developed within the private sector or from within the ranks of organized medicine itself. The creation of hospital and medical insurance plans that were professionally controlled and the emergence of commercial indemnity insurance plans represented a central development in health care in midcentury America. While there are commonalities between the FSA medical care programs and these broader historical developments, there are differences that set the FSA experience apart as well. In order to appreciate the similarities as well as the uniqueness of the FSA experience, a brief review of changes in the financing and delivery of medical services during the 1930s and 1940s is warranted.

The emergence and expansion of voluntary group prepayment insurance plans stands as perhaps the most important of these changes. By the end of the 1940s, voluntary health insurance was transforming the financing of American medicine. The majority of these plans came from the private sector, were tied to employment, and emphasized hospital insurance. These included commercial indemnity insurance plans, Blue Cross hospital insurance plans, and physicians' service bureaus. With few exceptions, these plans did little to alter traditional forms of medical practice and, in spite of the concerns raised by organized medicine, they posed little threat to the growing dominance of the medical profession. Instead, they were marketed as a means of financing medical and hospital care and served to reinforce the hegemony of the medical profession.

One of the best-known voluntary health insurance programs emerged from the Texas heartland when some 1,200 Dallas schoolteachers were offered group hospitalization coverage by Baylor University Hospital in 1929. The idea took root and expanded to other states, evolving into the hospital insurance program known as Blue Cross. Blue Cross plans offered subscribers "first day, first dollar" coverage for the costs of hospitalization, but steered clear of covering medical services for fear of invoking a response from organized medicine. By 1940, thirty-nine Blue Cross plans existed, with a combined subscribership of over 6 million members. Within ten years, every state had a Blue Cross plan and enrollment exceeded 31 million.[20]

Taking their cue from the popularity and success of the Blue Cross program, medical societies began to develop their own voluntary insurance plans toward the end of the 1930s. These physicians' service bureaus (or medical service bureaus as they are also known) emerged state by state beginning in 1938 in Michigan and California. By 1946 there were 43 physicians' service bureaus with over 3 million members, the majority of which provided coverage for physicians' services in the hospital and represented a complementary development to the coverage provided by the Blue Cross plans. In time these plans evolved into what we know today as Blue Shield. They were one strategy through which organized medicine sought to limit the growth of insurance plans that were outside the direct control of the medical profession, including compulsory national health insurance.[21]

The AMA was slower to warm to the idea of voluntary health insurance than were individual medical societies at the state and county level. However, following an extraordinary series of events that are summarized in Chapter 2, the AMA eventually dropped its official opposition to third-party insurance in 1938, and the organization allowed state medical societies to set up their own medical service plans. By 1946, voluntary physician-controlled medical care plans were a critical element in organized medicine's shield against compulsory national health insurance.

While health insurance in the United States has historically been available as a benefit of employment, prior to World War II it was often only the largest employers who could afford to provide medical benefits of any kind to their work force. Alternatively, health insurance coverage was available in industries where strong union representation allowed workers to negotiate for health benefits or through small sickness indemnity plans available through guilds or a variety of social organizations. During World War II, labor shortages and a federally mandated wage freeze encouraged the growth and diffusion of health insurance as a benefit of employment. The explosive growth of third-party insurance in the postwar era consolidated this pattern.

Professionally controlled plans (Blue Cross and Blue Shield) and private commercial indemnity plans drew their membership largely from industrial workers and urban communities. As a result, third-party insurance marginally penetrated rural communities well into the 1950s. Moreover, these plans made few efforts to reach those low-income families that were the target of the FSA medical care programs.

A few innovative privately sponsored medical care delivery plans that stretched the boundaries of medical practice emerged during this period. Several of these plans operated in rural communities and used nontraditional financing and organization strategies, including the use of salaried medical staff, capitation, or consumer participation. Many were bitterly contested by organized medicine at the time. In this group one might include the Kaiser-Permanente plans, philanthropic demonstration projects, and consumer cooperatives, such as the Farmers' Union Hospital Association. Philosophically, these programs had more in common with the FSA medical care plans than did physicians' service bureaus or Blue Cross plans. For example, consumer cooperatives were predicated on the active participation of members in governance issues.

The health care program developed under the auspices of construction and shipyard magnate Henry J. Kaiser was both innovative and controversial. The first of these plans was created when Dr. Sidney Garfield set up a medical care plan for the 5,000 Kaiser employees working on the Grand Coulee dam. From 1935 through the end of World War II, the Kaiser program expanded to include employees (and their families) at numerous construction projects undertaken as part of a massive federal jobs creation program, and during the war years it expanded to the workers at the Kaiser shipyards. Kaiser built clinic and hospital facilities near their construction sites and staffed them with salaried health professionals. Workers covered by this plan received all their health care in these facilities. According to historian Rickey Hendricks, the Kaiser program represented "the world's largest private health care system, the first group health plan in the nation to fully incorporate prepayment, group practice, and substantial medical facilities on a large geographic scale."[22]

One of the few health care programs that specifically targeted rural families was the Farmers' Union Hospital Association cooperative in Elk City, Oklahoma. The idea of cooperatives was not new. As early as the late nineteenth century, small farmers in rural communities had occasionally banded together to share the costs of marketing and production as a strategy to gain economic leverage with railroads, retailers, and other middlemen who controlled agricultural prices and markets. Dr. Michael Shadid, an immigrant Lebanese physician who had practiced for nearly two decades in the community, decided to establish a medical care plan based on cooperative principles. Backed by the Oklahoma Farmers' Union, Shadid started the Elk City Farmers' Union Hospital Association in 1929. For a one-time fee of $50, farmers became shareholders in the cooperative association. After that, the plan provided comprehensive medical and hospital care at a cost of $25 per year. With timely financial and political support from the Oklahoma Farmers' Union, Shadid and the Elk City plan weathered a storm of protest from local physicians and organized medical groups across the state, as well as from the AMA. By 1939, membership in the cooperative stood at nearly 10,000.[23]

Such indigenous rural health plans, however, were few and far between. Moreover, they were quintessentially local efforts and provided no meaningful model for national health planning. Even the Kaiser-Permanente plans were geographically concentrated in the western states of Oregon, Washington, and California, and were located where large industrial facilities or construction projects, good wages, and plentiful workers made such plans actuarially sound.

The medical care plans sponsored by the Farm Security Administration evolved in this milieu and therefore shared important similarities with the private sector health care plans. For example, the FSA medical care plans incorporated the principles of prepaid group insurance, and most FSA plans were based on voluntary participation by providers and patients alike. On the other hand, there were signal differences that set the FSA medical care programs apart from their better-known contemporaries. The most important difference was that the FSA plans were government sponsored. Another distinguishing feature was that FSA leaders sought to alter the pattern of physicians' practices in order to improve the accessibility and quality of medical care in rural areas. Examples of this are found in the agency's piloting of plans that incorporated capitation and salaried physician staff and its insistence on practice guidelines in the area of maternal and child health. In addition, the FSA's rehabilitation mission led the agency to try to modify those conditions that predisposed to ill health. This goal was immaterial to the insurance plans emerging from the private sector, especially those sponsored by professional bodies. Among the agency's most defining features was its incorporation of health education and preventive health programs into all facets of its health care programs. This was exemplified by the instruction the FSA gave on nutrition and the agency's collaboration with other public health agencies around sanitation issues, such as building privies and installing screens on windows and doors.

The FSA's mission to improve the access of a low-income population to medical care distinguished it from the majority of its contemporaries as well. It would have been actuarially·unsound and poor business practice indeed for private health insurance plans to specifically target low-income families for enrollment, and for the most part they made few efforts to do so. Although the wartime economic boom fostered the vigorous expansion of the private insurance industry, giving a growing number of citizens access to medical care, millions of elderly and poor citizens with substantial health needs remained uncovered by any health insurance.

Over time, evolving attitudes among participating physicians resulted in a crucial reversal in the relationship of physicians to the FSA programs in terms both of control and of actual delivery of care. Many physicians lost interest in what had once been a mutually beneficial partnership. The war's positive effect on physicians' income, for example, lessened the financial incentive that had once countered their anxiety about federal involvement in medical care.

The willingness of FSA leaders to work with the medical profession in developing a comprehensive medical care program was one of the program's great strengths. The FSA programs offered physicians sufficient autonomy and income to foster a cooperative attitude. In order to expand medical services in as many rural communities as it could, the FSA also made a virtue of adjusting to local circumstances. Some historians have argued that this strategy inevitably led the agency to accommodate local power groups as well.[24] As time passed and circumstances changed, however, physicians used their influence first to constrain and then to undermine the medical care program. This was a central paradox in the entire FSA rehabilitation effort: the agency's flexibility allowed it to better achieve many of its goals, but it also left the program vulnerable to powerful local interests that were often resistant to the agency's reform agenda.

Organized medicine's eventual alliance with conservative political opponents of the New Deal, and doctors' suspicion that the FSA programs heralded more extensive reforms such as national health insurance, also conspired to end the FSA's incursion into health care delivery. Emboldened by the nation's economic growth during the war and the inattention of a president focused on world events, these opponents whittled away at the FSA's rehabilitation mandate. Exacerbating this situation was the fact that after 1942 FSA medical leaders allied themselves with reformers within the government who favored compulsory national health insurance. The American Medical Association seized upon this relationship and raised the specter that the FSA programs represented the proverbial "entering wedge" with which national health insurance would be forced on the medical profession and society.

I feel that historians have been overly harsh in the criticism of Roosevelt and other New Deal politicians for their failure to pass national health insurance legislation. The emergencies the president faced during the Depression and World War II were enormous, and the opposition was formidable. National health insurance was only one of the programs on the New Deal agenda, and the determination with which it could be supported had to be balanced against the overall economic recovery program and later the war effort.

New Deal politicians did not enact legislation making health care reliably available to all Americans, but they did produce programs—notably the FSA medical care one—that significantly increased the availability of medical care to a large number of people who previously had had little access to it. These programs demonstrated the viability of innovative financing techniques and suggested ways in which the delivery of medical care could be improved. Without their example, the demand for health care that they exerted, and the political pressure exerted by their leaders and supporters, it is doubtful that voluntary insurance would have grown as it did and when it did. The forces supporting the status quo eventually gained ascendancy, and the FSA programs were terminated. Although the innovations they introduced did not, as agency leaders hoped, form a model for a national plan, aspects of these programs can be seen in major health care reforms of the past fifty years, both in this country and abroad.

ORGANIZATION OF THE BOOK

In order to understand the context within which the FSA medical care programs evolved, it is first necessary to understand some of the social, economic, and political realities of America in the 1930s. Chapter 1 provides a selective overview of medicine and health care during the Great Depression and a glimpse of the social situation that led to Roosevelt's first inaugural victory and the coming of the New Deal. This chapter summarizes New Deal rural relief and rehabilitation policy, beginning with the Federal Emergency Relief Administration (1933–35) and continuing through the Resettlement Administration (1935–37) to the creation of the FSA in 1937. It also briefly considers the statewide plan the agency established in North Dakota and South Dakota. Chapter 1 ends with the FSA ready to embark on a major expansion of the fledgling health care program started by the Resettlement Administration.

The FSA developed three major health care delivery programs, which emerged in roughly chronological order: the medical care cooperative program, first begun in 1935 under the Resettlement Administration and expanded by the FSA after 1937; the migrant health plans that the RA started in 1936; and the experimental health plans begun in 1942, just after the nation's entry into World War II. The agency also sponsored medical care plans in its highly controversial resettlement communities and a statewide plan in the two Dakotas, the farmers mutual aid corporations.

Chapter 2 provides a full description of the medical care cooperative program as it expanded under the FSA and briefly considers the medical care plans in the FSA's resettlement communities. Chapter 3 reviews the growth and development of the migrant health programs. Chapter 4 presents the novel experimental health plans.

Woven into the narrative of each chapter is the story of the mutable relationship that developed between the FSA and the medical profession, including local physicians, their organized representatives in county and state medical societies, and the American Medical Association. This analysis provides a detailed picture of the complex attitudes of local physicians toward the government's rural health initiatives between 1935 and 1942. It shows that many of the factors that led to the demise of the FSA medical care program can be traced to problems that antedated the nation's involvement in the war. In many counties, the establishment of medical plans during this period was prevented or delayed. In others, the programs operated, but falteringly. These problems reflected physicians' concern over the government's role in health care delivery.

World War II, called by one New Deal historian the "agent of the unraveling of so many liberal plans," had a powerfully deleterious impact on the social agenda of the New Deal.[25] Its influence on the FSA and its medical care programs in the period from 1942 to 1946 is fully covered in Chapter 5. This chapter also discusses the alliance that developed between conservative farm organizations, organized medicine, and congressional opponents of the FSA during the twilight of the New Deal, an alliance that was central to the demise of the FSA medical care programs.

The fate of the programs was also deeply influenced by the reinvigorated debate over national health insurance during the 1940s. This is the focus of Chapter 6. For many years, FSA leaders were cautious in their public statements on national health insurance for fear of alienating physicians whom they relied upon to provide services to their clients. In the years following the United States' entry into the war, however, FSA leaders became visible participants in the controversy over national health insurance. Their alliance with the proponents of national health insurance alienated many of their physician supporters and contributed to the termination of the agency.

The early care that FSA medical leaders took to promote their program as a temporary and local response to an economic crisis has been misinterpreted by historians as indicating that the FSA did not consider its medical care programs to be a template for broader health reform. In his 1968 history of the national health insurance movement, *The Lost Reform,* Daniel Hirshfield argued that FSA leaders never intended their programs to lay the groundwork for a compulsory system of national health insurance. Hirshfield does acknowledge that the FSA programs had widespread and lasting effects, such as lowering barriers to medical care among low-income rural families and influencing the popularity of private voluntary health insurance plans in rural communities in the postwar period. More recently, Rickey Hendricks in her history of Kaiser-Permanente notes that the "minuscule" FSA programs were not intended to be the basis for national health planning.[26]

Substantial historical evidence suggests that this was not the case. The individuals who directed FSA medical care viewed their efforts as an opportunity to explore the advantages and problems of federal tax-assisted medical care delivery. The experimental health plans, in particular, were designed as experiments in broadening federal involvement in rural health care delivery. In addition, the FSA's attempts to change the way in which medicine was practiced; its extensive preventive medicine efforts; and its systematic employment of nurses, nutritionists, and public health medical officers, show that agency leaders had more in mind than a temporary program for medically indigent farmers. Their intent was acknowledged by Michael M. Davis, an influential medical economist and social reformer, who in 1938 complimented the FSA for its pioneering effort. "Considering the attitude of the AMA," Davis wrote, "this is a real achievement . . . The extension of these medical care programs will set up a most important pattern for our whole rural population."[27]

In presenting the history of a program that has been relegated to a historical footnote, one has to guard against overstating its significance and influence. States, townships, and municipalities have assumed responsibility for the poor and sick through public support of pesthouses, almshouses, asylums, and hospitals since the colonial period. The establishment of quarantine laws in the nation's largest port cities and the growth of local and state health departments throughout the nineteenth century are additional examples of policies that placed authority over public health issues in government hands.[28]

The twentieth century, too, has been witness to periodic debate over the proper role of government in health and social welfare. During the Progressive Era, a broad political alliance, led by the Association for the Advancement of Labor, successfully promoted a reform agenda that culminated in the enactment of the nation's first unemployment laws, pure food and drug regulations, and worker's compensation statutes. Many states seriously debated tax-supported health insurance programs and three (California, New York, and Massachusetts) nearly passed compulsory state-based plans in the mid-1920s. The decision by the federal government to offer hospitalization coverage to veterans in 1924 extended a commitment initiated in the early nineteenth century with the Marine Hospital Service. Historical precedents for the involvement of governmental agencies in health and welfare activities at the municipal, state, and federal levels are plentiful and obviously relevant.

Nevertheless, it is clear that the FSA programs have direct ties to a number of innovative and well-known health care delivery programs in the postwar period. These include the Canadian health care system, health care plans developed by the United Mine Workers of America and the United Auto Workers, the community health center movement of the 1960s, and the health maintenance movement of the 1970s and 1980s. The epilogue charts a series of paths leading from the FSA to some of these programs. While each of these relationships deserves further exploration, for the time being I will leave this task to others.

CHAPTER ONE

Puttin' Out for the Farmers

The Evolution of New Deal Rural Rehabilitation Policy

Unemployment was about 27 percent and everybody was sort of at the same level. And the way you might describe it was that they were sort of dulled. There wasn't much hope . . . There was nothing.
THOMAS GARLAND MOORE JR., 1979

It was in mid-September of 1929 that the stock market began its historic ten-week slide, culminating in the infamous "Black Tuesday" of October 29, when the market lost a quarter of its value in a single day. Over the next few months, stock and commodities markets lost nearly half their value, industrial production declined by 9 percent, and imports declined 20 percent. Two years later, the gross national product was down 29 percent and industrial output was at half its 1929 volume; construction alone declined by nearly 80 percent. Between 1929 and 1931, the collapse in the banking and investment industries was nearly total as economic investment plummeted 98 percent and an average of two hundred banks went bankrupt each month. Unable to recover their savings from failing banks, many citizens simply stopped buying consumer products, further compounding the paralysis in the business community. Wage levels fell rapidly, and unemployment soared to an unprecedented 25 percent.[1]

By the time Franklin Roosevelt took the oath of office in March 1933, the United States was mired in an economic depression whose effects reverberated throughout the industrialized world. Few sectors of the nation's economy were unaffected by the Depression, but in rural communities it had the added effect of magnifying longstanding problems. An understanding of the Depression's acute impact on rural America as well as the hard-core poverty of many rural communities is central to understanding the evolution of the New Deal rural rehabilitation programs, including the health programs sponsored by the Farm Security Administration. This chapter reviews the social and economic milieu within which the FSA medical care plans emerged, focusing on the three key participants: low-income farmers, rural physicians, and the federal government.

A BELT OF SICKNESS AND MISERY:
RURAL AMERICA IN THE DEPRESSION

For rural Americans, the Great Depression added acute financial insult to longstanding economic injury. Beginning in the late nineteenth century, farm value and farm ownership went into a steady decline, and the reality of farming came into sharper and sharper contrast with the mythology of agriculture as a noble and salubrious calling. As Roosevelt's Secretary of Agriculture Henry Wallace once remarked, farms were areas of "vicious, ill-tempered soil with a not very good house, inadequate barns, makeshift machinery, happenstance stock, tired, over-worked men and women—and all the pests and bucolic plagues that nature has evolved."[2]

Farm mechanization and land consolidation reshaped rural life to an unprecedented degree in the years following the Civil War as larger and more prosperous landowners increasingly dominated the agricultural marketplace. Buying even a single tractor allowed large farms to replace the equivalent of four farm hands. The concentration of wealth, technology, and land into fewer and fewer farms steadily eroded the position of the family farmer in American agriculture. Between 1900 and 1930, these forces pushed nearly 25 percent of the nation's farmers into tenant farming or work as itinerant farm hands. Even during the 1920s, a period of blissful prosperity for much of the nation, many of those employed in agriculture saw only the continued decline of their economic position.[3]

In some regions of the country, farm life had been so hard for so long that the Depression itself went virtually unnoticed. "The Depression was not a noticeable phenomenon in the poorest state in the Union," Mississippi writer Eudora Welty observed.[4] Per capita income in the southern states was half the national average. For example, in 1929 mean per capita income in the United States was $650 per year. In the predominantly rural state of South Carolina, however, income averaged just $270 per year. In that state's rural townships, per capita annual income was a mere $129, one-fifth the national average.[5]

Based on available measures of health status, rural communities fared worse than towns and cities during the period of the Depression. The prevalence of preventable or treatable diseases such as pellagra, hookworm, syphilis, tuberculosis, malaria, and typhoid were stark reminders that rural America had been materially poorer than the rest of the nation for decades. Moreover, rural communities had a more difficult time generating the public funds needed to build the infrastructure that would have made improved access to medical services more meaningful, such as better roads, public health programs, or even hospitals.

Although the relationship between socioeconomic status and health had been articulated before the Depression, the depth and breadth of the economic collapse brought this link into dramatic relief. Lack of money to pay for medical services forced many people to go without them, and a single serious illness was enough to plunge a large and steadily increasing percentage of American families into seemingly irretrievable indebtedness. Even if a family called in the doctor, follow-up visits or recommended therapy was too often a financial impossibility: "I have to treat many families, shutting my eyes to the fact that not one of my instructions can be carried out," lamented one physician.[6] "Signs are accumulating that the American people are not going scot-free from damages to health in this depression," U.S. Public Health Service statistician Edgar Sydenstricker noted in 1934. "Recent inquiries are revealing not only a higher sickness rate among the poor as compared with those more favorably situated, but also a higher rate among those who have suffered most marked economic challenge during the past three or four years."[7] The same link between rural poverty and health was made by sociologists Arthur Raper and I. D. Reid in their 1941 text, *Sharecroppers*: "It is generally known that about half the South's people are unable to pay for medical care, that there are over 1,500,000 cases of malaria in seven Southern states, that hookworm and pellagra are most common in areas where tenancy is highest, income lowest, and educational rating poorest."[8] Statistics such as these provide substance to Welty's comment that the South was "a belt of sickness, misery and unnecessary death."[9]

REPORT OF THE COMMITTEE ON THE COSTS OF MEDICAL CARE

In 1927, a group of prominent American public figures and institutions were brought together through the support of the U.S. Public Health Service and eight philanthropies to form the Committee on the Costs of Medical Care (CCMC). Made up of representatives from medicine, the social sciences, public health, and private philanthropy, the CCMC had as one of its main goals the development of recommendations that would ensure adequate health care to all Americans at a cost that was within their means to pay. The statistics and narratives gathered by the CCMC ruthlessly compared the health conditions of underserved populations with those of the general population. The CCMC began its research in the still-prosperous 1920s and it was well under way by the time of the 1929 crash. Between 1930 and 1932, the CCMC published twenty-seven volumes filled with a wealth of detail on the health status of Americans and on the problems of the nation's health care system. Amid the deepening Depression in 1932, the CCMC published a summary of its findings, entitled *Medical Care for the American People: The Final Report of the Committee on the Costs of Medical Care*.[10]

The CCMC took as its starting point that medicine was a social good. It

57-year-old sharecropper woman. She has black beads hung between her breasts as a remedy for heart disease. Hinds County, Miss., June 1937. Dorothea Lange.

concluded that access to medical care and cost were the primary determinants of health status for underserved populations. The CCMC found that socioeconomic status determined both the quality and availability of medical care, although it also noted that inadequate care was the norm for most Americans regardless of wealth. According to the CCMC, medical resources were plentiful enough, but they were not "distributed according to needs, but rather according to real or supposed ability of patients to pay for services."[11] For example, New York, a relatively wealthy and urbanized state, boasted one physician for every 621 individuals, while a much less affluent and more rural state, such as South Carolina, had just one doctor for every 1,431 inhabitants. In extremely poor rural counties, physician-to-population ratios sometimes exceeded 1 to 20,000, and many rural areas had no practicing allopathic physicians at all.[12]

The CCMC also warned against the growing dominance of specialty practices, supported the growth of group (as opposed to solo) practice, and called for a concerted effort to increase the number of general practitioners and for the reinvigoration of their financial and professional status. Although later reformers drew upon the CCMC's data and conclusions when advocating national health insurance, most CCMC members did not support such a measure. Rather, the *Final Report* recommended the growth of voluntary group health insurance as a means of evenly distributing the financial burden of medical care and improving the access of all citizens to the care they needed.

Although 39 of the 49 members on the committee signed off on the *Final Report*, there were dissenters on both the right and left, and several minority reports were issued. One, submitted by eight private practitioners, rejected the summary recommendations as too radical and a prelude to compulsory health insurance. The AMA subsequently endorsed this position, referring to the *Final Report* as an "incitement to revolution."[13] Reformers on the committee, on the other hand, took the opposite view. They worried that voluntary health insurance would become an obstacle to compulsory national health insurance and would leave the poor without adequate medical care. Two individuals, including Edgar Sydenstricker, refused to sign the document because its recommendations were so incremental and did not challenge the medical profession's dominance over the practice of medicine. The divisions within the CCMC over its summary proposals, and the dissatisfaction that the report engendered on both sides of the political spectrum, foreshadowed the debate over national health insurance that unfolded over the next two decades. Still, many of the CCMC's recommendations were progressive by the standards of the time, and subsequently were incorporated to varying degrees in a number of the private, professional, and government-sponsored health reforms that emerged during this period, including the FSA medical care programs. As Milton Roemer, one-time assistant chief medical officer for the FSA, later recalled: "By the time Farm Security got started, the idea of insurance as a way of taking care of illness which is unpredictable in the individual but is predictable in the group was in the air in liberal circles."[14]

BEANS ARE WINNING OUT:
MIGRANT HEALTH DURING THE DEPRESSION

Deterioration in the quality and security of farm life, deepening poverty, and declining health status were problems all across rural America. As wrenching as the Depression was for many rural families, however, its impact fell disproportionately on those whose tie to the land was most tenuous. For decades, hundreds of thousands of low-income farmers, tenant farmers, and share-croppers scraped a meager living out of lands depleted by years of single–cash cropping. As marginal as their livelihood had been, those who lost their farms altogether and went in search of more arable land or more stable employment faced an even bleaker predicament.

Displaced rural families swelled the ranks of the American migrant labor force throughout the decade. The transition from subsistence to homeless penury could be rapid and utterly complete. "They lost everything," T. G. Moore remembered, "and they lost it in five or six years."[15] Even more well-to-do farmers were bankrupted and found themselves without their homes or their livelihoods. Journalist Paul Taylor, traveling the Southwest with his wife, FSA photographer Dorothea Lange, captured this devolution in the personal history offered by a laconic and once prosperous Texas cotton farmer:

1927—made $7,000.
1928—broke even.
1929—went in the hole.
1930—still deeper.
1931—lost everything.
1932—hit the road.[16]

Natural disasters, such as drought, floods, and dust storms, compounded the distressed condition of rural America and directly contributed to the greatest internal migration in our nation's history. A terrible and lasting drought began in 1932, and over the next few years, thousands of farms were first desiccated and then buried under a succession of spectacular dust storms. The prolonged dry spell did not discriminate according to wealth and landholding, but the least economically resilient farmers were most deeply affected. Nationally, perhaps as many as 650,000 families worked land classified as submarginal and they abandoned their smothered farms in droves.[17] The impact of the drought on the health of rural communities was as dramatic as the dust storms themselves. Between February and April 1935, the Kansas weather bureau reported a total of only thirteen dust-free days. Over that same period, infant mortality increased by 66 percent and death from respiratory infections tripled.[18]

What Nature failed to reclaim was likely to be foreclosed upon by cash-

Farmer and his sons walking in the face of a dust storm. Cimarron County, Okla., April 1936. Arthur Rothstein.

starved local banks. The national banking crisis that accompanied the Depression had a severe impact on agriculture. As a result of the economic downturn in agriculture in the previous decade, farm indebtedness grew from $8.5 billion to $9.1 billion. As prices declined, individual farmers tried to maintain their incomes (and pay their debts) by increasing production. They could not suspend the law of supply and demand, however. Agricultural surpluses piled up, and the price of farm produce was driven to historic lows.

With farmers unable to make debt payments, farm foreclosures became a regular occurrence in rural communities during the Depression. Violent protests, such as the 1932 Sioux City milk strike, sometimes erupted as frustrated farmers attempted to force the government to take action to raise farm prices. In some communities, auctioneers, bankers, and law enforcement officials were threatened, beaten, or, in one notorious incident, killed when they attempted to auction off farms. Angry farmers were themselves subject to violent retribution by local law enforcement officers and in several farm states, National Guard units were called out to quell community protests. These dramatic events, coupled with the obvious distress of the agricultural sector, caught the eye of politicians in Washington and helped generate support for Roosevelt's farm program during his first administration.[19]

The combined effect of drought, decades-long agricultural trends, and the Depression was that there were more and more people to accommodate on less and less arable land. Overall estimates of the number of families displaced was over a million, and from 1935 to 1938, nearly 100,000 persons a year crossed into California alone.[20] The majority of these internal refugees migrated from the dust bowl states of Texas, Oklahoma, and Arkansas, or the parched prairies of the Dakotas. These dispossessed families radically altered the demographic composition of the nation's migrant population. Prior to the Depression, the typical migrant worker in the South was black, while those in the West were predominantly of Mexican or Asian descent.[21] By the mid-1930s, 85 percent of migrants were displaced white Americans. This fact focused unprecedented attention on migrant welfare issues. The plight of the country's new migrants was considered a national disgrace and received enormous political, literary, and social attention over the next decade.[22]

The appalling living standards of displaced families was attested to by their median net income (in 1933) of $110 and an annual period of employment that averaged 23 weeks.[23] Migrants crowded into thousands of squalid squatter villages, known as Hoovervilles or Little Oklahomas, on the periphery of agricultural communities. Others lived in private grower-owned camps where conditions were only marginally better. Even had they been so inclined, local and state governments lacked adequate resources to inspect these camps, and statutes defining employer responsibility and housing and sanitation codes went unenforced. A few county health departments or individuals within public and private welfare agencies were sympathetic to the migrants, but most deferred to local growers, who were hostile to regulatory oversight and little disposed to provide more than meager accommodations to transient workers.

Historically, migrants have fared worse than other rural groups in terms of baseline health status and access to medical care, housing, and sanitation.[24] This was never more apparent than during the Great Depression. A study by the California state health department found that the status of migrant children across a range of health measures fell significantly below that of a comparison group of nontransient rural children. This study blamed poor housing, inadequate nutrition, and nearly nonexistent sanitation facilities for most of the migrant children's sicknesses. For example, high rates of anemia and vitamin deficiency in migrant families were an indication of chronic malnourishment. Infant mortality rates soared during picking season in counties where migrants settled.[25]

Illnesses already prevalent among migrants further increased during the Depression. Hidalgo County, Texas, a community with a large number of mi-

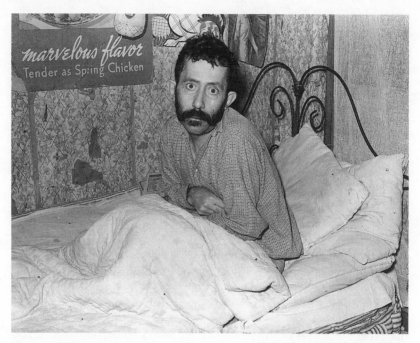

Mexican with an advanced case of tuberculosis. He was in bed at home with other members of the family sleeping and living in the same room. Crystal City, Tex., March 1939. Russell Lee.

grant workers, had a tuberculosis rate that was eight times higher than the national average during the decade of the 1930s, averaging nearly 400 cases per 1,000 residents compared with 51 cases per 1,000 nationally. Death from tuberculosis among migrant families was nearly two hundred times higher than among U.S. residents as a whole. In the sugar cane region of Louisiana's bootheel, malaria was twenty times more frequent, typhoid was nine times more frequent, tuberculosis was twice as common, and mortality from gastroenteritis and dysentery in migrant children was eighteen times more frequent than it was in the rest of the state. Separated from their families during the harvest season, male migrant laborers contracted sexually transmitted diseases, such as syphilis and gonorrhea, at high rates. Pellagra, a preventable disease caused by a lack of the vitamin niacin, caused the death of 645 migrants in Texas in 1938. While the Depression worsened the physical and material health of migrant workers in general, ethnic minorities and immigrant families were hit hardest. African Americans, as a group among the lowest wage earners in the South, had significantly higher infant mortality rates and experienced higher rates of most infectious and chronic diseases. Mexican Americans harvesting citrus fruit in California and Chinese immigrants picking beets in the Pacific Northwest fared equally poorly. [26]

Even without the Depression's insuperable demands on local and state resources, the majority of migrants were ineligible for local, county, or state assistance. Stringent residency criteria required individuals to have established permanent residence for one to three years before becoming eligible for benefits. Local school systems, too, were off limits to nonresident farm worker children. Even when residency requirements were met, children were essential breadwinners for many migrant families and often attended school only when it did not interfere with crop harvesting. "Education is in competition with beans in this country," observed journalist Carey McWilliams, "and the beans are winning out." [27]

Rural communities were often ambivalent in their attitudes toward migrants. Deeply felt prejudices against migrant workers clashed with need for their labor when there were crops to be picked. "There is a popular habit of calling persons 'workers' when they are needed to harvest a ripened crop," the *Los Angeles Times* editorialized, "and of referring to them as 'bums' during the slack season." [28] The more common sentiment was expressed by the *San Diego Sun*: "The only time a bum is expected to come to California is when we need him as a harvest hand. What right has he to come between seasons?" [29]

Concern for public health often cloaked prejudice against migrant workers and their families. Without basic amenities, such as sanitation and safe drinking water, migrant camps were ripe environments for epidemic disease, and contagion was an understandable concern for many rural communities. [30] This legitimate fear was also manipulated by health officials and community leaders to justify acts of violence against migrants. Systematic campaigns to destroy squatter villages were frequently led or inspired by local health departments under the banner of protecting the public's health. One local public health officer was quoted as saying that migrants were "a people whose cultural and environmental background is so bad that . . . no advances have been made in living conditions among them and ethically they are . . . far removed from the desire to attain the privileges which present day culture and environment offers." [31] Newspapers and politicians alike railed against the rising migrant tide. "MIGRANT HORDES INVADE KERN," a local newspaper headline ran in Kern County, California. The *Los Angeles Herald* called migrants a "moron element infecting this county," and "a big danger to respectable citizens." [32]

PLOW BEAMS AND FRIED CHICKEN:
THE DEPRESSION'S IMPACT ON RURAL PHYSICIANS

Physicians as a group fared better than most Americans during the early years of the Great Depression, but it was not unheard of for physicians to be on relief. According to one research monograph, there were approximately 600 physicians on relief in 1935, only 0.3 percent of all doctors. However, while doctors generally avoided bread lines, nationally their annual incomes declined between 20 and 40 percent from 1929 to 1932.[33] Moreover, the economic impact of the Depression on doctors was not evenly distributed. As the Committee on the Costs of Medical Care documented, even before the Depression, the pockets of rural general practitioners were significantly more shallow that those of their urban and specialist colleagues. According to the CCMC, general practitioners made less than half the money specialists did and rural general practitioners made less than their town counterparts. In the best of times, collection rates for rural physicians hovered around 75 percent of billings. During the early years of the Depression, however, collections rarely exceeded 40 percent and at times fell below 20 percent of billings. In the drought-ridden states of Oklahoma, Texas, and Arkansas, physicians' incomes plummeted 50 percent.[34] William Wright, a general practitioner from Morgan County, Kentucky, recalled the frustrations of a typical country doctor of that era: "The doctor did the best he could. Sometimes you got paid, sometimes you didn't get paid. Sometimes you'd get a bartered deal. They'd bring in food or they'd work for you. That kind of thing. There wasn't a whole lot of money and there wasn't a whole lot of care."[35]

Organized medicine, eager to maintain the tradition that care of the poor was a local or individual responsibility, pointed proudly to the historic willingness of the profession to provide charity care, or accept lower fees or barter when patients were unable to pay their doctor's bills. Well-to-do physicians (usually urban specialists) were more capable of absorbing the costs of free care, either by charging higher fees to paying patients, or by writing their services off as an educational experience or simply charity. However, overburdened and underpaid rural doctors were hard pressed by the growing demand for free care. "I have been paid for 'medical services rendered' in molasses, hogs, chickens, eggs, corn, potatoes, ham, beans, all kinds of vegetables," one rural physician recalled. "The biggest item I ever accepted in payment for medical service was 200 plow beams. I still have some of those plow beams."[36] T. G. Moore recounted a conversation he had with a local doctor: "Doc Johnson said to me one day, ' I've eaten just about all the fried chicken and black eyed peas I can stand.' And I know it to be a fact, a lot of his pay was in just that, you know."[37] As time passed, doctors were left with limited choices. They could refuse to continue what was basically free service or else they had to accept less reimbursement and more fried chicken and plow beams.

ALTOGETHER A DIFFERENT PROPOSITION:
THE NEW DEAL COMES TO THE FARMER

The economic collapse of the 1930s overpowered the ability of private welfare agencies, local charities, and community physicians to cope with the health needs of their communities. While for some groups these needs were acute conditions arising out of the Depression itself, for others chronic deprivation was merely intensified by the catastrophic economic downturn. In either case, the need for relief persisted and worsened from 1929 until 1933. This was the state of the nation when Franklin Delano Roosevelt promised a "New Deal" for an anxious and struggling people.

Roosevelt was elected president in November 1932 and assumed office the following March. Public expectations ran high, and Congress was unusually compliant as the new president and his aides launched the New Deal. Through a legislative barrage and the liberal use of executive orders, the tone and direction of the Roosevelt domestic agenda was established during the first hundred days of his presidency. Although historical opinions on the achievements and limitations of Roosevelt and the New Deal vary widely, few would question that Roosevelt redefined the role of the executive branch and the responsibilities of the federal government in American society.[38]

The president brought with him to Washington a core group of academics and professionals to advise him on domestic policy. The group, which became known as the Brain Trust, included Rexford Guy Tugwell, who was named undersecretary of agriculture. Tugwell had taught agricultural economics at Columbia University. He had become enamored of the collective approach to farming promoted during the 1920s in the Soviet Union, and he believed fervently in government activism to remedy social ills. Throughout his career in government, his strong views embroiled him in highly visible confrontations with American business leaders, Congress, and the conservative agricultural establishment. As secretary of agriculture, Roosevelt selected Henry Wallace. Wallace came from a distinguished farm family in Iowa and was one of the nation's foremost agricultural authorities. Frequently, his intimate understanding of farm issues put him at odds with ivory-tower reformers such as Tugwell. Wallace was deeply affected by the abysmal conditions he witnessed in many rural communities, but as an agricultural scientist he was less confident than Tugwell about the effectiveness of government intervention as a cure for endemic rural poverty. Both Wallace and Tugwell were key figures in the events that led up to the formation of the Farm Security Administration and its medical care programs.[39]

It is germane to our story that following Roosevelt's election, the American Medical Association hastily called a meeting of its house of delegates to define its position on government intervention in the practice of medicine. Or-

ganized medicine feared a rapid expansion of the federal government in medicine as part of the president's promised introduction of sweeping welfare legislation. The AMA hoped that its guidelines would channel reform efforts in
a direction more in keeping with the interests of the medical profession. The
guidelines adopted in 1934 by the AMA stated that there should be no third-
party intermediary in the doctor-patient relationship and that control over all
aspects of medical practice was the sole prerogative of the medical profession.
If the government were to have any role—and the AMA was not enthusiastic
about the idea—it should be strictly limited to paying the medical bills of the
indigent, provided determination of eligibility and provision of service were
left in the hands of local authorities, including physicians.[40] While there were
shifts in organized medicine's posture over time, these principles proved
durable and survived reasonably intact in the face of repeated challenges by
the national health movement over the next fifteen years.

On March 21, 1933, President Roosevelt sent an omnibus relief measure
to Congress, creating the Federal Emergency Relief Administration. Roosevelt chose Harry Hopkins, a close advisor and friend, to head the massive
agency. Hopkins had been a social worker in New York City, where he worked
for various agencies, including city health agencies and the Red Cross. He became known as a skillful administrator who got things done, even if it meant
bending or ignoring rules. When the Depression first began, then-Governor
Roosevelt chose Hopkins to run the Temporary Emergency Relief Administration that had been created to deal with the state's unemployment and relief programs. After Roosevelt's election as president, he brought Hopkins to
Washington. The Federal Emergency Relief Administration was obviously
modeled on Hopkins' work in New York. Upon assuming the directorship,
Hopkins was quoted as saying "the average amount of relief per family in the
U.S. has been about fifty cents a day per family. Well, this is a fine how-do-
you-do to fourteen million people living on fifty cents a day. It is ridiculous and
absurd. It can't be tolerated."[41]

The FERA was principally a direct relief agency intended to meet the broad
welfare needs of the nation's poor, irrespective of region, occupation, or race.
In line with Roosevelt's sense of the political importance of state sovereignty,
an Emergency Relief Administration was set up in each state to channel federal money as well as the matching state funds required by the legislation. Relief to eligible individuals and families usually came in the form of outright financial aid; loans and credits for delinquent bills or rents; and food, shelter,
clothing, fuel, and other household supplies. Sometimes work relief in the
form of low-paying or part-time jobs was available to the unemployed.[42]

Information collected by government agencies throughout the Depression
showed that the burden of illness on families on relief was enormous. Statistics collected by the U.S. Public Health Service during the early years of the

Depression are particularly telling. With funding from the FERA, the Public Health Service, in conjunction with other federal and state agencies, conducted a health inventory in some ninety cities and twenty-one states beginning in 1933. In 1935, it published a report showing that families on relief had 47 percent more acute illness and 87 percent more chronic illness than families with incomes over $3,000 per year. In figures strikingly similar to those reported by the Committee on the Costs of Medical Care, the survey indicated that a third of all serious disabling illnesses among families on relief went without physician care. In the next higher income group, nearly a third also received no physician care.[43]

More sobering still was the mounting evidence that infant mortality rates in many states were rising for the first time in decades. It was already recognized in academic and public health circles that infant mortality was inversely related to income. Since half of all babies were being born to families on relief, the implications were ominous. By the 1930s, medical and public health experts agreed that medical advances had the potential to ameliorate the effects of poverty and that infant deaths were largely preventable. Improvements in prenatal care, physician-supervised delivery (especially in a hospital), and early postnatal care would benefit mothers, infants, and society itself. This conviction would find direct expression in the health education and medical services in the FSA programs and other New Deal social welfare programs.

Administrators within the FERA realized that deteriorating conditions in rural communities required a more encompassing strategy than direct monetary support. Within a year, the FERA created a separate division devoted to rural relief and rehabilitation, the Division of Rural Rehabilitation and Stranded Populations. Another early modification of FERA policy dealt with medical relief, which was not initially included as a reimbursable expense under FERA direct relief guidelines. On September 10, 1933, the FERA issued Rules and Regulations Number 7, qualifying medical care for relief funds. So long as treating physicians set their fees "at an appreciable reduction from the prevailing minimum charges," the FERA provided money to pay medical bills.[44] The Division of Rural Rehabilitation and Standard Populations and this regulation set the federal government on a path that led to the creation of a nationwide medical care delivery program over the next decade.

The involvement of the FERA in medical care was a natural extension of its relief duties. The FERA helped fill the gap created when "the entire system of medical care for the needy and medically needy—poorly organized, uncoordinated, relying primarily on local government resources and the charitable services of physicians, voluntary health and welfare agencies—collapsed."[45] Families were initially paid directly by FERA and in turn they were expected to make good on their unpaid doctor bills. This was consistent with established FERA policy and also minimized potential conflict with organized medicine, which remained wary about intermediaries, fiscal or otherwise, in the doctor-patient relationship.

Physicians' sentiments regarding the federal role in negotiating and paying for discounted medical services varied. Both the AMA and individual practitioners recognized that local charities and local and state governments did not have the funds to provide for the medically indigent and that increased federal support for health care to this group was probably necessary. Commentary in medical journals of the day indicates that individual physicians and medical societies were also acutely aware of their humanitarian obligations in the face of historically unique circumstances. The Missouri State Medical Association, for example, stated that the AMA had urged cash-starved physicians to participate in the FERA plans because the "common aim, as stated by the administration, is the provision of good medical service at low cost, to the benefit of the indigent patient and the physician, nurse, dentist, and taxpayer. It is hoped that physicians will enter heartily into the spirit of these rules."[46] In 1933, the *Journal of the Missouri Medical Association* observed that while the FERA program undoubtedly represented "political regulation," most physicians and organized medical groups would accept the government payments because the "primary consideration of the medical practitioner is the welfare of the people, the preservation of their health and their care in sickness."[47]

The Federal Emergency Relief Administration's approach to medical relief did not technically violate the principles adopted by the AMA following Roosevelt's election. The program did not require physicians to participate, and it did not prescribe what treatment physicians should offer. Moreover, except for its discounted fee schedule, it did not disrupt traditional fee-for-service medicine. Nevertheless, in some quarters of the medical community, misgivings were voiced that the aid program could prove a dangerous precedent. Even supporters of the FERA program articulated a strong preference for initiatives, preferably from the medical profession itself, to resolve whatever medical care problems existed. This is a recurring theme that is woven throughout this history.

The payment mechanism initially adopted by the FERA also caused a good deal of dissatisfaction among both doctors and patients. Physicians worried that government payments to families for medical bills might be diverted to pay for other family expenses. Furthermore, the FERA required that individuals or families seeking money for medical care apply to local welfare or state ERA offices for approval for medical care already received, or for assurance that their bill would be paid once care was obtained. Sometimes doctors were unwilling to provide care until prior approval had been given, and many patients waited to call in the doctor until they had the needed authorization, resulting in more costly care and greater morbidity. Delays in payment or preapproval were common, reflecting a process that was at times cumbersome, bureaucratic, and subject to common prejudices. "One black man did come in for health care and he had to go to the county judge then to get approval," remembered T. G. Moore. "The judge turned him down and he died that night in the mule barn. The next day (it was the first time I'd seen anything like it) the entire black community was standing across the street silently and did not move."[48]

As timid as the FERA's intervention into medical care was, however, it offers an early indication that Roosevelt and his domestic advisors were willing to involve the government in what had traditionally been the dominion of the medical profession. For example, fee schedules averaged 50 percent of the established rate and were negotiated between the government and county and state medical societies. Not only did the FERA wrestle with physicians' groups over reduced fee schedules for families on relief, but when controversies over reimbursement emerged, they were referred to the state emergency relief administration (ERA) for resolution. The negotiation of the federal government with physicians' representatives over fees and fee disputes signaled an important, if subtle, shift in the relationship between the federal government and the medical profession. Given the economic state of affairs and the failure of nonfederal relief agencies, the government had the opportunity, will, and power to begin to influence matters that had heretofore been unchallenged. This minor incursion quickly assumed greater force when the New Deal moved from direct relief to economic and social rehabilitation beginning in 1935.

New Deal welfare programs, of which the Federal Emergency Relief Administration was one of the first, forced the nation to confront the fact that many Americans lived in chronic poverty. These programs, along with the visual and verbal images of the Depression carried in popular magazines such as *Life*, the *Saturday Evening Post*, and *Harper's Weekly*, made it impossible for American society to ignore the destitution that had been hidden for so long in the hollows of Appalachia or the hills of Arkansas. The deep poverty of the South, in particular, came as a genuine shock to many who flocked to Washington to be a part of the New Deal. Eleanor Roosevelt became a tireless advocate on behalf of that region's poor, especially African Americans, who made up a disproportionate part of the sharecropping and tenant farming population. Throughout her husband's presidency, the first lady steadfastly supported the New Deal's domestic agenda and helped directly on such programs as the Tennessee Valley Authority, the Rural Electrification Administration, the Appalachian Regional Commission, and the FSA resettlement programs to be described later.[49]

As the months passed, Roosevelt's liberal agricultural advisors came to realize that it was a defective social fabric, not the Depression alone, that was at the core of the South's poverty. The foundation of southern society was agriculture, and the cement for this foundation was its farm tenancy system. In

1933, Charles S. Johnson, a leading black sociologist at Fisk University, published a scathing critique of farm tenancy and the adverse socioeconomic impact that single cash cropping (usually cotton) had for the majority of the region's farmers. The "general depression," wrote Johnson, "reached the South when it was already prostrate and sadly crippled by an outworn tenancy system."[50] The awareness that they faced a structural economic cancer brought into sharper focus by a temporal economic crisis provided FERA's administrators with the direction the agency and its New Deal successors would take in their rural relief and rehabilitation efforts throughout the country.

Compounding these structural problems were unanticipated consequences arising from New Deal legislation itself, most notably the Agricultural Adjustment Act (AAA). Congress passed the AAA in March 1933 to stem the downward spiral in agricultural prices. The AAA readjusted existing farm debt, extended credit to stricken farmers, removed cropland from active production, and provided direct and work relief for some of the nation's poorest farmers. The AAA also stabilized the agricultural market by raising the price paid by the government for agricultural goods, by reducing the exchange rate on the dollar (thereby boosting the income from agricultural exports), and by imposing tariffs on imported farm products. The combination of price supports and protectionism had its desired impact: agricultural prices began to rise soon after the act was passed.[51]

The AAA, however, exacted a terrible human price for its success in shoring up farm prices. When landowners were offered money to leave a certain percentage of their land unplanted, they naturally started with their least productive acreage, which was invariably farmed by their tenants. Although they were specifically instructed under AAA policy to maintain the same number of tenants on their land when removing cropland from production, many simply evicted their tenants. Furthermore, AAA policies that supported farm prices, eased agricultural credit, and modernized farming preferentially benefited larger (and politically well-connected) producers at the expense of the small producer. Forced out of farming, hundreds of thousands of additional farm families joined the growing wave of displaced Americans entering the migrant work force.[52]

The human wreckage caused by the AAA generated concern in the Roosevelt administration, but it was especially devastating to Tugwell and his supporters. They believed that the only way to redress the structural repression represented by farm tenancy was to promote farm ownership. To that end, they urged Secretary of Agriculture Henry Wallace to convince the president of the need to consolidate rural relief programs into a single agency whose sole purpose was to provide for those farmers falling between the cracks of the AAA.

While Roosevelt was sympathetic to Tugwell's concerns, he was acutely aware that such a move might alienate southern Democrats whose support he needed to pursue his legislative agenda. In February 1934, the president finally asked Hopkins to begin a rehabilitation program for low-income farmers under the auspices of the Federal Emergency Relief Administration. The Division of Rural Rehabilitation and Stranded Populations created by Hopkins provided loans to farmers to buy their own land and the technical supervision to improve the efficiency of their farming. The new program was orchestrated by a large, decentralized field staff. Many of these field staff were former social workers or teachers who believed that education, technical advice, and financial aid would improve the socioeconomic conditions of the country's most vulnerable farm groups.[53]

Friendly supervision and easy credit became the hallmarks of all subsequent New Deal rural relief and rehabilitation efforts. Low-income farmers applied to the local ERA supervisor for credit to buy seed, equipment, and animals and were in turn provided with technical advice by the government representative. In effect, the federal government became the surrogate landlord for thousands of farmers, sharecroppers, and day laborers. Alabama sharecropper Nate Shaw remembered this transition well. Turning his back on his old landlords who had cheated him in the past, Shaw recalled: "I wound up with all of 'em. I went onto the government and the government furnished me. The news was out through the settlement—the federal people was in Beaufort puttin' out for the farmers." It was, added Shaw, "altogether a different proposition."[54]

In emphasizing credit and technical supervision, the Federal Emergency Relief Administration's rural rehabilitation programs followed longstanding Department of Agriculture approaches embodied to an important degree in the activities of the Agricultural Extension Service and the Bureau of Agricultural Economics (BAE). What made the program distinct was its focus on the rural poor and its goal of promoting economic self-sufficiency and competitiveness in a group that many agricultural economists and certainly larger growers considered anachronistic in the modern agricultural economy.

By the spring of 1935, the nation's economy was still languishing, and Roosevelt was increasingly anxious about his government's massive commitment to direct relief. Roosevelt's advisors, in particular Harry Hopkins, urged the president to move away from "the dole," and Roosevelt agreed. The president substituted a series of social welfare programs that emphasized work relief and rehabilitation in place of programs like the FERA. As part of this transition, Roosevelt acceded to the demands of administration militants, such as Tugwell, who felt that more should be done for America's rural communities. On April 30, 1935, the president created the Resettlement Administration (RA) by executive order, and elevated the agency to cabinet status. The RA assumed the rural relief and rehabilitation effort begun by the FERA, including its medical care program. Roosevelt appointed Tugwell to head the RA, a decision that ensured that the agency would be a political lightning rod. The RA, like the FERA, concentrated its initial efforts on the South, since that region was perceived as having the most significant social and economic problems.[55] However, the agency had national reach, involving itself in North and South Dakota when they were hit by drought and providing some assistance to migrant workers in California, Texas, and Florida. The RA's primary mission was to enable destitute and economically distressed farm families to achieve a measure of economic independence. The RA was far more ambitious in its goals than the FERA had ever been, and armed with "more plans than funds," Tugwell began a spirited attack on rural poverty.[56]

The Resettlement Administration pursued its rural rehabilitation mandate through a bewildering series of programs that frequently overlapped and thus gave ammunition to the agency's critics. The RA bought land from landowners and then provided credit to tenants, sharecroppers, and day laborers to buy this land. The RA also provided financial and technical assistance to individual farmers and farming cooperatives through its standard rehabilitation loan program. The agency funded approximately 150 resettlement projects and four suburban or "greenbelt" communities around the country in which displaced rural families, and even homeless urban dwellers, were given housing, a small plot of land for personal use, and technical advice on cooperative farming and marketing.[57]

Resettlement Administration field staff also encouraged their clients to establish farming, equipment-buying, and marketing cooperatives. Cooperatives were a manifestation of agency leaders' belief that small family farmers could compete with larger farmers and that they gave these traditionally disenfranchised citizens experience in participatory democracy. They were a major means by which Tugwell and others who believed in the potential of the rehabilitation programs to improve the socioeconomic status of low-income rural families hoped to reach their goal. As T. G. Moore recalled, "The long term goal was to become a part of the community, to be stable, to be secure, with roots down. To have the kids off at school, to be able to worship as they saw fit, and to be able to intermingle on a par with other people, not as a second-class citizen . . . As far as I'm concerned that's what it was all about."[58]

Established farm interests, in particular the American Farm Bureau Federation (AFBF) and the Grange, were deeply distrustful of the thrust of the New Deal's latest rural initiative. The consolidation of federal rural rehabilitation programs into a cabinet-level agency under Tugwell's control and the RA's broad and ideologically driven attack on rural poverty did not go unchallenged by these groups, who had a vested interest in preserving their influence over

Farm Security Administration supervisors with a rehabilitation client. Greene County, Ga., June 1939. Marion Post Wolcott.

rural communities and agricultural policy in the United States. According to a Memphis paper, Roosevelt's action had placed the New Deal rural rehabilitation effort "into the hands and under the blight of social gainers, do-gooders, bleeding hearts and long-hairs who make a career of helping others for a price and according to their own peculiar, screwball ideas."[59]

Several of the programs instituted by the Resettlement Administration were highly controversial and embroiled first the RA and then its successor, the Farm Security Administration, in a political maelstrom from which agency leaders could never quite escape. The resettlement of farmers onto more productive land and outright land redistribution in particular were anathema to the AFBF and the Grange, and led to charges of "sovietism" and "communism." The agricultural establishment saw the promotion of farming cooperatives in order to make small farmers more competitive economically and the RA's emphasis on participatory democracy as a direct threat to their traditional influence. Traditionalists in the Department of Agriculture, including many in the Agricultural Adjustment Administration and the Agricultural Extension Service, disliked Tugwell intensely, and derided the undersecretary and his supporters as "boys with their hair ablaze," or, worse still, "urbanites."[60]

THE GOVERNMENT HANGS OUT ITS SHINGLE:
EARLY NEW DEAL RURAL HEALTH INITIATIVES

Nascent rural health initiatives emerged under the auspices of several federal agencies during the early years of the Roosevelt administration. The Department of the Interior, the U.S. Forest Service, and the Department of Agriculture all claimed some authority over rural welfare matters, health care
included. This reflected the widespread calamity facing rural communities,
the dispersion of rural welfare responsibility among a handful of federal agencies, and the often noted fragmentation characteristic of the New Deal. In the
spring of 1935, for example, the Department of the Interior began considering ways to include medical care as part of its subsistence homestead program.
In May of 1935, Charles Pynchon, director of the Division of Subsistence
Homesteads, sent a letter to Olin West, secretary of the AMA, calling West's
attention to the "acute problem of rural health and rural medical care."[61] In
his letter, Pynchon solicited West's opinion on how best to proceed to deal with
the very difficult problem that had arisen: "I am writing this to ask you whether
out of your knowledge and experience you would have any suggestions to make
to us . . . one of the methods which has been suggested is that of the so-called
health insurance option." Pynchon received a curt response from West in
which it was suggested that the government stay out of medical care delivery
altogether. "I am strongly of the opinion that it will be difficult to secure good
medical service on a contract basis," wrote West, "you will probably find such
an arrangement to be unsatisfactory to everyone involved."[62]

Two months later, as part of the consolidation of federal rural relief and rehabilitation programs, the Department of the Interior turned over control of
the subsistence homestead program to the Resettlement Administration. RA
leaders felt a growing concern about the poor health of their rural clients. The
strenuous demands of farming were often proving unmanageable for families
whose emotional and physical reserves were sapped by chronic and acute illnesses. Any plan that improved the ability of rehabilitation borrowers to obtain needed medical care would have the dual impact of improving clients' rehabilitation potential and protecting government loans.

Resettlement Administration leaders also approached the AMA regarding
their concern about the inaccessibility of rural medical care and its negative
impact on the agency's rehabilitation goals. The voluminous data compiled by
the Committee on the Costs of Medical Care and the health survey by the
U.S. Public Health Service notwithstanding, the AMA was unconvinced that
any special effort was required. In a 1936 article on "Rural Medical Service,"
Morris Fishbein, the influential editor of the *Journal of the American Medical
Association*, wrote that "these counties are receiving all the medical care that
they demand and considerably more than they can pay for. No evidence has

been found to show that the people who live in these counties have complained of any lack of services."[63] The AMA did not reject the government's request out of hand, however. Instead, the AMA suggested that the RA broach the subject with state and county medical societies. This became the standard approach by which the medical care programs developed under the umbrella of the New Deal rural rehabilitation programs, first under the RA and then the FSA.

There were several reasons for the AMA's response to the RA request. First, there was significant pressure both within and outside of organized medicine to be more flexible regarding the medical care of indigent families during the Depression. Second, the economic hardship of many practicing physicians helped AMA leaders to be reasonably accommodating in regard to the RA proposal. A 1933 editorial in the *New England Journal of Medicine*, hardly a bastion of medical sedition, took the AMA to task for ignoring the problems facing the medical community and society: "What is happening to the medical profession is the absence of its own planning to meet the exigencies of a radically changing society . . . Is it too much to hope that the influence of our powerful national organization will be on the side of the overburdened physician . . . when he is suffering the worst exploitation he has ever known?"[64]

Finally, the politics of national health policy affected the AMA's strategic decision. Organized medicine was well aware that some of Roosevelt's advisors were advocating national health insurance. By leaving the ultimate decision to its member societies, therefore, the AMA gave the appearance of cooperating with the government while distancing itself from any specific endorsement of health insurance, or federal medical care programs.

Franklin Delano Roosevelt's promise of a New Deal for America included the creation of a social welfare safety net. Soon after he took office, Roosevelt convened the Committee on Economic Security (CES) and charged the group with crafting the legislation that resulted in the 1935 Social Security Act. The CES considered national health insurance, along with old-age pensions and unemployment and disability insurance, in its early Social Security proposals. A small working group formed by the CES developed a health insurance plan as part of the larger social welfare bill. This group included several of the most vocal advocates of compulsory national health insurance within the administration, such as I. S. Falk and Edgar Sydenstricker, both of whom had been involved with the CCMC. Roosevelt was unwilling to risk the legislation by pushing for the health insurance provisions, and the final bill was considerably more modest in the health-related activities it funded. Titles V, VI, and VII of the Social Security Act provided money to support medically indigent and crippled children, for maternal and child health programs, and for federal and state-based public health programs.[65]

In the wake of his decision to drop health insurance from the Social Security bill, Roosevelt created an executive branch committee, the Interdepartmental Committee to Coordinate Health and Welfare Activities. As its name suggests, the committee was charged initially with coordinating the various health programs that were at that point scattered through a number of federal bureaucracies and with monitoring the health-related provisions of the Social Security Act. During Roosevelt's second term, the committee's mission expanded to include an assessment of the nation's health needs and consideration of a national health program. But that is getting ahead of our story. For now, the important point is that the committee's formation showed the government's growing interest in medical care, which alarmed organized medicine and provided the context in which Tugwell and his staff would pursue the issue of medical care for rehabilitation clients.

After receiving indications that it could broach the subject of rural health and medical care with local and state societies from the AMA, the RA quickly initiated discussions with several state medical societies, most of them in the South. In line with the agency's cooperative philosophy, local relief and rehabilitation supervisors worked with client families and local physicians to create medical care cooperatives. Typically, an RA representative approached area physicians for support for a prepaid group medical care plan funded through the federal rehabilitation loans of participating families. Thus, the pattern for the development of the rural health cooperative program began under the Resettlement Administration, and this was expanded once the FSA was created. Meanwhile, a more immediate role for the Resettlement Administration emerged in response to an acute public health crisis in North and South Dakota.

CRISIS IN THE NORTHERN PLAINS:
THE FARMERS MUTUAL AID CORPORATIONS

Drought and economic depression devastated the northern plains during the early 1930s. In 1932, 60 percent of North Dakota's farms were sold to pay off creditors of the state's destitute farmers. For two years, the Federal Emergency Relief Administration provided a measure of relief to the ravaged region. However, Roosevelt's decision to terminate FERA funding in December 1935 precipitated a welfare crisis of catastrophic proportions. In North Dakota alone, 60,000 families, a third of the state's entire population, were dependent on county welfare boards. Between the two states, half the population was on some form of relief and virtually everyone else qualified.[66]

Health conditions deteriorated rapidly and growing alarm over the situation prompted a meeting between local, state, and federal officials, the last including the U.S. Public Health Service, the Works Progress Administration (WPA), the U.S. Children's Bureau, and the RA. The two states' medical so-

A bleached skull on this parched, overgrazed land gives warning that here is a land which the desert threatens to claim. Pennington, S.D., May 1936. Arthur Rothstein.

cieties, pharmacists, and hospital associations participated in the discussions as well. Surgeon General Thomas Parran himself traveled to the region to facilitate negotiations between organized medical groups in both states and federal health and welfare agencies.[67]

Discussions began in August 1936, and over the next three months all parties agreed to coordinate their activities in order to halt the precipitous decline in health conditions in the devastated states. By October, local, state, and federal agencies, with the acquiescence of the North Dakota Medical Association, launched a novel public-private program to provide medical care coverage to families on relief: the farmers mutual aid corporations. A nearly identical statewide plan was started six months later in South Dakota.

The AMA was not a full partner in the negotiations, but its influence is obvious in the final form given to the two farmers mutual aid corporations. Both plans generally adhered to the AMA's 1934 guidelines, for example. Through the farmers mutual aid corporations, families on relief were able to obtain medical care at home or in a doctor's office, hospitalization coverage, major surgery, emergency dental care, nursing care, and drugs. The agreements stipulated that families apply to county or state welfare agencies for authorization to become members. Once approved, they were issued medical cards that they presented to participating doctors or hospitals in order to receive medical care.[68]

Financing of the farmers mutual aid corporations began with a modified fee-for-service mechanism, and later evolved into a form of capitation. In the former, physicians billed families using a drastically discounted fee schedule. In the latter, a predetermined amount of money per enrolled individual or family was set aside from which physicians would be directly reimbursed for medical care. The fee schedules adopted by the farmers mutual aid corporations were 30 percent less than the usual minimum fees charged by physicians, an indication of the financial hardship facing the professional community there. For example, surgical procedures were paid at a flat rate of $50.

In addition to acute medical services, each plan covered preventive services, such as immunizations, and prenatal and postnatal care for mothers and infants. In an early sign that the RA's interest went beyond creating a mechanism to pay for medical services, the farmers mutual aid corporations set specific obstetrical standards for participating physicians. This included "an agreed minimum number of prenatal visits, delivery in the home or hospital, and necessary post-natal care."[69] The incorporation of preventive health practices, particularly those relating to maternal and child health, was one of the defining characteristics of the New Deal's rural health effort. The effect these policies had on the health of farm families and the anxiety they kindled in organized medicine are considered in more detail in later chapters.

The impact of the medical care programs in the Dakotas was immediate. Increasing access to medical care for farmers and ensuring at least partial compensation to physicians resulted in a surge in the use of medical services in both states. Farmers mutual aid corporation members were hospitalized more frequently and received more medical services of all types than noneligible families. Based on the available evidence, these differences cannot be fully attributed to more severe baseline health conditions among plan members.[70]

The magnitude of the crisis in North and South Dakota forced AMA spokesmen to restrain their public commentary on the farmers mutual aid corporations. Pressure from both state governments and organized medical groups in the two states, where as many as a third of all physicians were "not making a living and some are in actual want" also held the AMA in check.[71] The farmers mutual aid corporations, wrote the *Journal of the American Medical Association*, were "unobjectionable" so long as they remained emergency measures limited to families on relief.[72] The journal emphasized that the government had assured the AMA that the federally subsidized plans would "distribute case income among the physicians in an equitable manner" and that they would not "break down the much-desired family doctor relationship."[73]

In private, the AMA hierarchy harbored grave doubts about developments in the two states. Inside the RA, it was widely acknowledged that "every effort was made by the A.M.A. to prevent the organization of the corporation and to defeat its purposes."[74] The contractual relationship between physicians and the federal government, codified in the articles of incorporation, was of special concern to the AMA. According to the *Journal of the American Medical Association*, the articles of incorporation were overly broad and vague. Consequently, they gave the government "powers far beyond those necessary to enable them to carry on the activities for which they were supposedly created," and allowed the Farmers Mutual Aid Corporation to last "far beyond any period within which specific farm relief in the states named can be regarded as likely to be needed."[75]

The leaders of organized medicine feared that this experiment might be a blueprint for continued federal intervention in medical care delivery even after the crisis in the region passed. In 1937, the AMA warned the state medical societies in North and South Dakota of the "dangers inherent in these excesses," but both groups refused to reconsider their participation, even when specifically asked to do so by the national organization.[76] The cooperation of North and South Dakota physicians with the federal government also generated harsh criticism from colleagues in other states. The criticism focused on the contract between physicians and the federal government, the low fee schedule, and physicians' fears that the plans could interfere with free choice of doctors.

The doctors in North Dakota and South Dakota reminded their critics that the medical care needs of some 310,000 people on relief were simply too much for local and state welfare boards and physicians to handle. Dr. H. A. Brandes wrote an article for the September issue of the *Journal of the Iowa State Medical Association* summarizing the reasons that led him and his colleagues to work with the Resettlement Administration.

> Realizing it was necessary for us to take immediate action to get federal funds into the state to provide for our people and to give assistance to our physicians, especially in the smaller communities, we decided to take our chances with a cooperative rather than with the individual client. Of course, we do not approve of medical cooperative organizations but under the present arrangement the physician . . . will be paid for his services.[77]

Brandes clearly shared some of the concerns raised by the AMA and his colleagues in other states, openly wondering if programs beginning with indigent farmers in the Dakotas might end by advancing "the cause of state or socialized medicine." He went on to state:

> The articles are drafted along broad lines and if carried out would prove vicious and far-reaching in their effect on the practice of medicine . . . [W]e have been

assured that it is not the intention of the RA to exercise the powers granted in the Articles . . . for the purpose of initiating a new form of medical practice. In this we may be entirely wrong in our premises . . . There are some who may criticize us for not acting wisely but we feel we are at least helping our people and our doctors through a hard winter.[78]

Brandes's comments capture the conflicting emotions many rural physicians experienced as they considered whether to cooperate with the Resettlement Administration and later with the Farm Security Administration. The financial strain many rural doctors experienced and their concern for their struggling neighbors made it difficult for them to stand in the way of a public program that provided needed medical care. At the same time, they hoped that such measures were temporary and did not portend graver developments.

The tension experienced by practicing doctors was felt much less keenly, if at all, by the leaders of organized medicine. Although the AMA portrayed itself as the defender of private practice, the organization was dominated by well-to-do urban specialists who had little in common with the majority of rural general practitioners. Neither the RA nor the FSA ever received the sanction of the AMA for their medical care programs. The AMA's behind-the-scenes maneuvering during the Dakota crisis, and its hands-off response to the RA's urgent request for assistance in easing the acute health problems in rural areas, indicate that the association was hostile to federal health programs of any kind, even those targeted explicitly to the rural indigent.

The government's relationship with Dakota physicians and state and local medical societies was a volatile one. Early commentary in the *Journal Lancet* (put out jointly by the medical societies of the two states) indicates that the North Dakota Medical Association viewed the medical care program plan as a humane response to a crisis and appreciated the fact that it provided needed income to struggling rural doctors. In the first year of the program, doctors in North Dakota and South Dakota received 53 percent and 61 percent of their billings, respectively. The government's policy of working with allopathic physicians, its reliance on general practitioners to provide most of the care, and its inclusion of minimal standards of care for certain types of practice (especially maternal and child health) all reinforced the status of the medical profession and were crucial in eliciting physician support.[79]

The problems raised by physicians in both states revolved mainly around limitations on reimbursement, eligibility criteria, and confusion as to the comprehensiveness of medical services available to plan members. Although each of these issues had been specifically addressed in negotiations with the two state medical societies prior to the creation of the farmers mutual aid corporations, some physicians still complained. South Dakota physicians were apparently less enamored of the federal program than their colleagues to the north. When escalating program costs and unexpectedly large demand for services led the government to institute a $1 monthly membership fee, the decision rankled participating physicians in both states and contributed to a growing disaffection among South Dakota physicians and county medical societies. "The FMAC appears to us to have failed, not only the Medical Profession which entered into agreement with [it] in good faith, but also the public," the *Journal Lancet* noted. "We are pleased to learn that a few of the societies have taken a positive position against renewing further agreements."[80] Despite receiving 60 percent of their billings, opposition to prorating and antagonism over the monthly membership fee grew more vocal in South Dakota, eventually prompting the organization to temporarily abrogate its agreement in 1938.

Issues of professional control also surfaced. For example, South Dakota doctors took pains to emphasize that it was the state medical society, not the federal government, that was behind the development of the Farmers Mutual Aid Corporation. Organized medicine's desire to maintain sole control over medical practice was an ongoing issue as the government's rural medical care delivery program unfolded.[81]

Whatever anxieties physicians felt, they were not sufficient to derail the health programs in North Dakota and South Dakota completely, nor would they do so when the Farm Security Administration expanded the medical care program it inherited from the Resettlement Administration. The RA lasted just two years before it was terminated and its rural programs were transferred to the FSA. Although the Resettlement Administration itself was short lived, it was responsible for placing the problems in rural health and medical care within the purview of the federal government's rural rehabilitation effort. In a 1937 *Interim Report of the Resettlement Administration*, submitted to Congress just before the agency was abolished, Tugwell noted that the eight medical care cooperatives were "very promising" but "still not adequate."[82] It was left to the Farm Security Administration to make medical care delivery a cardinal feature of the New Deal's rural rehabilitation programs.

CHAPTER TWO

A Matter of Good Business
The FSA Medical Care Cooperatives

*And the way that program worked was that I went down and talked to the
doctors and believe it or not, I got some cooperation.*
THOMAS GARLAND MOORE JR., 1979

By the fall of 1936, Tugwell had become a political liability for the Resettlement Administration and for President Roosevelt. At midnight on December
31, 1936, two carefully orchestrated changes in the RA took effect. First, Tugwell relinquished his post as administrator and was replaced by his deputy administrator, William Alexander. Second, an executive order demoted the
agency from cabinet status and made it a part of the U.S. Department of Agriculture. Over the next six months, opponents and supporters of the Resettlement Administration fought over controversial programs such as resettlement
communities, land redistribution, and cooperatives, continuing battles that
had kept federal farm policy in turmoil for years. The institutional tumult
finally simmered down in July 1937 when Roosevelt signed into law the
Bankhead-Jones Act. Bankhead-Jones was a compromise bill, passed largely
because of concern over farmers' agitation. It made three primary provisions
for the rural poor: loans to tenants to enable them to buy land, rural rehabilitation loans for the purchase of seed and farm equipment, and programs to
stabilize submarginal land use. Soon after it was passed, Secretary of Agriculture Wallace replaced the Resettlement Administration with a new agency in
the Department of Agriculture, the Farm Security Administration. There was
more continuity than change as a consequence of this reorganization. The
FSA continued all of the RA programs, including those that involved resettlement, and it retained the RA's decentralized field staff. The agency's new
name, however, reflected an effort on Wallace's part to deemphasize the controversial resettlement concept and to underscore the goal of preserving the
small family farm through such traditional Department of Agriculture approaches as loans and technical education.

FSA field staff continued to be friendly and helpful to clients, but the agency expected its loans to be repaid. FSA Administrator Will Alexander knew that loan repayment was one measure by which the agency would be judged in Congress. This political reality soon clashed with the agency's mission of rehabilitating the nation's poorest rural families. For example, the need to show a healthy balance sheet led some local rehabilitation supervisors to preferentially enroll those farmers most likely to make good on their loans, rejecting loan applications from the most marginal farmers. Because African Americans had fewer resources and less education than their white counterparts, and because of bias on the part of some local supervisors, as a group they found it more difficult to obtain FSA loans. This problem led to a certain amount of self-flagellation by the FSA's liberal staff and left the agency open to periodic criticism from more radical farm organizations, such as the Farmers Union and the Southern Tenant Farmers Union, as well as from later historians.[1] On the plus side, however, the agency's efforts to ensure loan repayment contributed to what became one of the New Deal's most ambitious and successful rural rehabilitation efforts: the FSA medical care programs.

Building on the model started by the Resettlement Administration, the FSA significantly broadened the government's involvement in rural health care. By 1942, there were medical care cooperatives in a third of the rural counties in the United States, and the agency had a significant migrant health program as well. This chapter describes the growth and development of the FSA's medical care cooperatives. Looking at the development of this program, one can see the issues around which conflict between the government and rural doctors arose and how the FSA attempted to resolve or avoid these problems. The issues included doctors' concern about reimbursement and professional autonomy, and their suspicion that the FSA program was the opening gambit in the federal government's drive to enact some form of national health program.

DIGGING DEEPER:
MEDICAL CARE AND RURAL REHABILITATION

Will Alexander was an excellent choice to head up the new agency, and under his leadership the FSA became the New Deal's undisputed champion of low-income rural families. Born into a poor family in rural Arkansas, Alexander graduated from Vanderbilt University's Divinity School. He maintained a deeply felt affinity for the nation's destitute rural citizens and became one of the South's strongest proponents of racial equality. Called "Dr. Will" by his friends and "nigger lover" by his detractors, Alexander was one of the New Deal's most benevolent figures.[2] Between his impoverished rural upbringing and his liberal credentials, he had both the understanding and the will to implement the New Deal's commitment to improve the lives of low-income farmers, sharecroppers, and migrant workers.

Alexander's humanitarianism did not blind him to the political constraints of his position, and he gained a reputation as an adroit political operative. He expanded the programs started by the RA under Tugwell. Alexander understood that the rural rehabilitation programs had evolved in response to both an acute economic depression and long-standing socioeconomic conditions. Addressing both immediate and historical problems required creativity, flexibility, and the commitment of a large and dedicated field staff. These were the characteristics upon which the FSA built its rehabilitation programs. Paradoxically, however, the same characteristics also contributed to the more bedeviling problems the agency faced.

Alexander relied on a decentralized field staff to promote the FSA's rural rehabilitation programs, including the agency's health-related efforts. There were variations from state to state and from county to county in the makeup of the FSA medical and nonmedical field staff, although the key field officers were the relief and rehabilitation supervisors. These individuals provided technical guidance to farmers and assumed most of the responsibility for coordinating the development and administration of any given county medical care unit. Like T. G. Moore, many were recruited to work with the FSA from the communities in which the agency operated and many knew the local medical community personally. An unusually large number of field supervisors were drawn from the ranks of teachers and social workers, two professional groups with considerable interpersonal skills. Supporting roles were played by district, state, and regional staff employed by the FSA, including public health nurses, who often provided baseline physical assessments of FSA borrowers; home management supervisors, who provided nutritional and educational advice to farm women; health services specialists, individuals hired by the FSA with experience in medical care delivery, health education, or public health; and sanitary engineers, usually provided by state public health units or the U.S. Public Health Service.

Corresponding to these positions was an administrative hierarchy that paralleled that of the agency as a whole, beginning at the county level and progressing up to district, state, regional, and national offices of the FSA's Health Services Branch. In Washington, the FSA Health Services Branch was headed by Dr. Ralph C. Williams, a U.S. Public Health Service medical officer with considerable experience in rural health issues. Williams was assigned to the RA as its chief medical officer in January 1936, and formally assumed his duties in June in the midst of the Dakota crisis. Williams stayed on as chief medical officer when the FSA was created in 1937, and he recruited a fellow Public Health Service medical officer, Dr. Frederick Mott, to be his assistant in early January of that year. The FSA drew on the U.S. Public Health Service to staff most of its regional and district medical offices as well.[3]

Alexander was committed to the fledgling medical care programs and considered them a vital element of his agency's overall rehabilitation mandate. With his support and active encouragement, the FSA immediately stepped up the development of medical care cooperatives. In August 1937, Alexander wrote a letter to FSA regional directors asking them to "dig deeper" to reach the most disadvantaged families and requesting that they "assemble as much general and specific information as practicable that would support my contention that many of our clients have inadequate health care."[4] Alexander wanted real case histories to justify the expansion of the medical care effort to Secretary Wallace, the Congress, and the president.

The field staff responded quickly to Alexander's directive. A fusillade of detailed, often heart-rending accounts of poor health and inadequate access to medical care bombarded the national office. Helen Browning responded that many farmers endured their illnesses "matter-of-factly" and "what is usually considered 'ordinary medical needs' . . . are generally done without."[5] "I am not going to give full details of everyone of our rehabilitation families," wrote Renae Lamb, a field worker in the FSA Home Management Bureau, "but I can safely say that there is a health problem no matter how small in every one of our families."[6] Some of these stories found their way into the FSA's promotional literature. "Jim Black neglected his work all spring," one pamphlet states. "His neighbors said he was lazy because he was slow about his work. Two weeks ago, he was unable to get out of bed and the doctor was called. We found that Jim had a hernia. He hadn't talked about it because he was ashamed that he couldn't pay for an operation."[7]

Correspondence from the field offices to the national office cataloged "flagrant" and unmet health needs, such as tooth abscesses, eye ulcers, pellagra, tonsillitis, tumors, venereal diseases, and "female troubles," many of which required "immediate attention."[8] Letter after letter illustrated the point that chronic disease and a tradition of substandard health care were complicating the FSA's already difficult task of rehabilitation. Poor road conditions, lack of transportation, and few and distant medical practitioners only made matters worse. "The nearest licensed physicians are at Andrews, Murphy, and Franklin," wrote Humphrey Browning from North Carolina. "These are from 21 to 42 miles from Graham County clients . . . clients do not have cars or wagons for transportation."[9] When rural rehabilitation supervisors asked families directly if they would participate in a prepaid medical care plan, the majority said they would. According to T. G. Moore, his clients "were aware that they didn't have medical care and they were prepared to get it."[10]

The picture of the medical profession that emerges from these field letters is a mixed one. While many rural physicians were in dire economic straits themselves, their sense of duty was usually deeper than their bank accounts. Refusing to provide medical care when payment was likely to be bartered, late, or not forthcoming at all was difficult for doctors whose patients were usual-

ly neighbors. However, the letters from the FSA's field staff made it clear that
without some guarantee that the bills of FSA clients would be paid, a growing
number of rural doctors would be unwilling or unable to provide medical care.
"We must remember that these doctors have carried this burden so many years
prior to any government emergency program to such an extent that we can't
expect them to continue," wrote Kathryn Ware.[11]

The response to his memorandum confirmed for Alexander and his staff
that there was a pressing need for a coordinated and expanded response to the
medical care needs of the agency's clients. Excursions by FSA leaders from
the national and regional offices to the nation's "Tobacco Roads" substantiat-
ed the voluminous anecdotal evidence submitted by the field staff. By 1937,
few in the FSA disagreed with Kathryn Ware when she admonished, "Something should be done and done quickly."[12]

Compelling economic incentives for providing health care to FSA clients
came in the form of data indicating that 50 percent of loan defaults were attributable to sickness. Echoing a theme used by the RA in initiating its health
programs, FSA leaders often stressed the economic benefits of their rural
health plans in the context of the agency's overall rehabilitation mission. As
one FSA pamphlet put it: "A family in good health was a better credit risk than
a family in bad health. So far as the government was concerned the program
was simply a matter of good business."[13] Chief Medical Officer Williams told
the 1939 American Public Health Association Annual Convention the program was "an incidental by-product of a depression-born loan program for
farm families unable to obtain credit elsewhere, and designed to accommodate a special economic group only."[14] Accenting the economic advantages of
the medical care program helped pacify the FSA's critics and, for a while at
least, limited the ability of conservative farm groups and the AMA to derail the
program. For supporters of rural rehabilitation, the medical care programs
were a humanitarian necessity that had the added convenience of being economically sound.

The economic justification was genuine as well as politically expedient.
However, it does not fully explain the agency's experimentation with a range
of approaches in its effort to improve the health and economic well-being of
its clients. The scope and organization of the FSA medical care programs were
much broader and more innovative than any that had been previously attempted by the federal government and bespeak a commitment to govern-
ment-supported health care beyond the temporary, emergency measures the
agency purported to be providing.

THE FSA COMES A-COURTIN':
SETTING UP THE MEDICAL CARE COOPERATIVES

The FSA focused considerable attention on the South. The region's chronic poverty, its large landless black population, and the conviction among New Dealers that the region's socioeconomic woes derived from its dependence on farm tenancy and single cash cropping made it a prime target for the FSA. Analogous structural deficiencies existed in regard to health care. Because it was poorer, the South had fewer hospitals, fewer and variably trained physicians, and provided less public health support at the state or county level than other regions of the country. The poor baseline health status of many of the region's poorest citizens was another reason to expand the medical care program in the South.

Like the RA before it, the FSA directed its efforts at local and state medical societies when it decided to expand its medical cooperatives. A review of the negotiations between the FSA and the North Carolina Medical Society gives a good example of the strategies used by agency leaders to promote the program as well as some idea of the resistance they faced.

When it was first approached by the FSA, the response of the North Carolina Medical Society did not bode well for the prospects of the medical care program in that state. In July 1937, the executive council of the association adopted a resolution stating its "unequivocal opposition to any plan of Federal control over the medical care of indigent persons."[15] In response, the FSA's Chief Medical Officer Ralph Williams began a series of communications with officers of the state medical society to press the issue. In a letter to the secretary of the North Carolina Medical Society, Williams expressed interest in working out a "memorandum of understanding" between the two organizations that would enable rural rehabilitation supervisors to work directly with county medical societies to develop medical care cooperatives in selected counties. To reassure the medical society about its good intentions, Williams noted the FSA's successful courtship of medical societies in other states. He also emphasized that the program's objective was to restore farmers' self-sufficiency and that it was not simple charity.[16]

The chair of the medical society's Medical Economics Committee, Dr. R. M. Houser, was distrustful of the FSA initiative and frankly dismissive of the actions taken by other states. However, he did not reject the proposal out of hand. Further information was needed, Houser wrote to the chief medical officer, "before we can come to any sensible conclusions." Houser wanted to know the counties targeted by the FSA, whether FSA clients had financial resources other than through federal loans available to them, and finally whether there was "any available medical service for these clients other than through an arrangement with the Administration." Houser ended his letter by stating confidently that "you can and will furnish me with the desired information."[17] Houser's acerbic response typified the reception the FSA encountered repeatedly in the course of developing its medical care program, particularly by the leadership of organized medicine at the state and national level. Over and over again, the FSA had to contend with the medical profession's fierce individualism and organized medicine's skepticism about the magnitude of rural poverty and sickness. Above all, FSA leaders had to overcome organized medicine's anxiety about entering into any agreement with the federal government. Houser eventually agreed to bring the proposal before the next meeting of the Medical Economics Committee. As chair of the committee, it was Houser's prerogative to convene it. Apparently he was in no hurry to do so.

Several months of inaction passed. Finally, an exasperated Williams wrote directly to Wingate Johnson, president of the North Carolina Medical Society. "I have gained the impression," stated Williams, "that Doctor Houser is somewhat inclined to be a little dilatory in calling his Committee together to consider our proposal. From our standpoint there is some urgency about the matter, inasmuch as our farm loans are now being made for the coming year."[18] The FSA chief medical officer then went to Raleigh to confer with Johnson in person. In the meeting, Williams reiterated the agency's position that poor health and inadequate medical care hindered economic rehabilitation and that this was the agency's primary goal. In an obvious appeal to the still-prevalent view that only the "deserving" poor should receive assistance, Williams emphasized in all of his letters with the state medical society officers that "it is our purpose to rehabilitate them and assist them to again become self-sustaining members of the community. It is obvious therefore, that medical care is a vital factor in the rehabilitation of these families."[19]

Williams's personal appeal worked, and Johnson gave permission for the FSA to discuss the medical cooperative program with county medical associations in the state. At that time there were over sixty counties in North Carolina in which the FSA had a significant programmatic presence. Fortunately for the FSA field staff, which was going to have to hammer out arrangements in each of these counties, Johnson approved a model memorandum of understanding in December 1937 that could be used as the basis for negotiations with individual county medical societies. This was typical of the strategy used by the FSA to expand its medical care program in other states as well.

Once these statewide agreements were signed, local FSA personnel, sometimes accompanied by district or regional FSA Health Services staff, attended meetings of county medical societies, presented the medical care plan, and asked physicians for their support. Health Services specialists, who were specifically trained to communicate the purpose and organization of the medical care program, often helped local supervisors explain the FSA plans to communities and to county medical groups. Not uncommonly, they also helped resolve problems when they arose. Taking their cue from their leadership, FSA representatives stressed the need for adequate health care in the context of the FSA's overall rehabilitation goals. If the county society agreed to endorse the plan, a memorandum of understanding, usually quite similar to that worked out with the state medical society (if available), was drafted and signed.

One of the most important factors in gaining local support for the medical care program from both physicians and farmers was the openness of the FSA to local concerns. This flexibility was rooted in both pragmatic and philosophical considerations. In operating a decentralized cooperative venture, it was necessary that the FSA accommodate the interests and priorities of those participating at the local level. This was expressed by including in the memorandum of understanding clauses of particular concern to county medical societies and local farmers. In some geographically extended counties, for example, a surcharge of 50 cents per mile was added onto the regular fee of $2 per home visit.[20] Also, while the majority of medical care units covered general practitioners' services, local negotiations between the FSA, local farmers, and county medical societies determined if the plan would include dental, surgical, hospital, or pharmaceutical services.

Accommodation of local concerns was also an expression of the FSA's commitment to participatory democracy. The Health Services Branch took the position that the details of any given medical care unit were "a matter for local determination."[21] As Chief Medical Officer Williams informed one regional director, "It is not within the province of this office to attempt to direct any specific phase of the program."[22] In practice, this attitude often meant ceding substantial control to doctors. For example, setting fee schedules, policing physicians' behavior, monitoring their use, and billing were largely left to local physicians and county medical societies.

The growth of the FSA medical care program was also facilitated by the agency's creative use of informational brochures and educational materials for staff and clients. Indeed, the agency was adept at promoting (some argued propagandizing) its mission and programs using a wide range of mediums; these included printed materials, radio commentary, newspaper and magazine articles, film, and, of course, the well-known documentary photographs. The backbone of this effort was a highly effective Information Division, first started in 1935 by Rex Tugwell. As indicated earlier, the work of the photographers and writers it employed made an important contribution to the history of this period. Two critically acclaimed documentaries produced by the Information Division, *The Plow that Broke the Plains* and *The River,* have achieved the status of American film classics.[23]

Negotiations between the FSA and state and county medical societies inevitably introduced a delay of months or even years in organizing medical co-

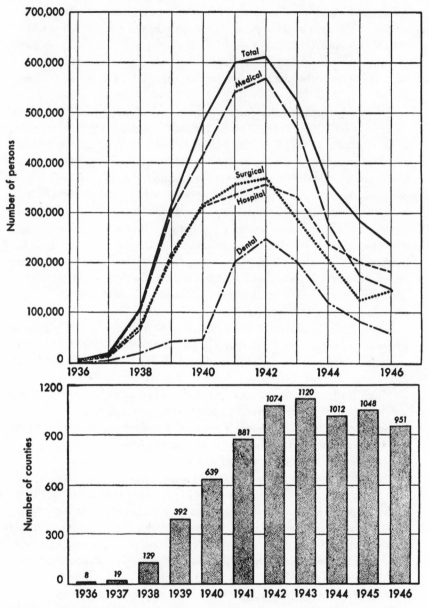

Farm Security Administration medical care plans: membership trends by type of service offered and number of counties with plans, 1936–1946. Membership data prior to 1940 represent estimates based on slightly incomplete reports. Temporary emergency programs in the Dakotas during this period are omitted. Frederick Dodge Mott and Milton I. Roemer, Rural Health and Medical Care. *New York: McGraw-Hill, 1948. Used with permission.*

Dr. S. A. Molloy examining Mrs. W. H. Willis and her family. Mr. Willis is a Farm Security Administration borrower. Blanch, N.C., October 1940. Marion Post Wolcott.

operatives. Nevertheless, intra-agency correspondence and the commentary winding through state medical journals indicate that the strategy of working with local physicians generally worked to the FSA's advantage. Although there were setbacks, as we shall see, the response from the medical community in most rural communities was largely positive. In 1939, the assistant director of Region IV, J. B. Slack, wrote to a local rural rehabilitation supervisor to inform him that "On Thursday October 5, the Medical Society of Rowen and Davis counties passed a resolution to cooperate with the FSA in organizing a Medical Care Program."[24] Reports filed monthly by the FSA's well-organized field staff kept agency administrators abreast of developments and provided a generally rosy perspective on the program's growth and development in North Carolina. An FSA supervisor from Weldon County wrote to his superiors in June 1939, "The Medical Society has passed a resolution making it possible, so far as they are concerned for any practicing physician to take care of a case of one of our members if called to do so."[25] Brief statements along the same lines peppered the *Journal of the North Carolina State Medical Society* for the next few years as the number of medical care units expanded from three plans in 1938 to thirty-eight by 1941.

A WORTHWHILE BEGINNING TOWARD BETTER HEALTH: MEDICAL CARE IN THE FSA COOPERATIVES

The general organization and policies of the medical care cooperative program were reasonably consistent across the country, although local variations were allowed and encouraged by the FSA. Medical care plans were based on the group prepayment insurance principle, a concept that was novel in rural communities at the time. All FSA borrowers and their families were eligible for membership. Typically an incorporated health association, often referred to as a medical care unit, was created by the memorandum of understanding and membership dues were paid into this entity. Dues averaged between $15 and $30 per year and were budgeted out of clients' rehabilitation loans with the help of the FSA supervisory staff. Less commonly, dues were paid out of pocket based on anticipated annual farm income. Membership fees were based on a family's size, its ability to pay, and the types of services negotiated with the county medical association. A respected community member, such as a local judge, banker, or attorney, usually served as the bonded trustee for the cooperative. In some communities the local FSA supervisor filled this role.[26]

Group prepayment created an insurance pool to lessen the cost of sickness to individuals by distributing the total cost of medical care over a larger population group. The common fund established was divided into twelve monthly allotments from which fees were paid to participating physicians. Less commonly, local FSA medical care units were based on a capitated system in which local physicians agreed to provide medical care to a family for a set amount. Perhaps as many as 100 medical care plans for rehabilitation families were based on a capitated model. Capitation was also used successfully in some of the resettlement communities, and in one of the experimental health plans the FSA created in 1942.[27]

Physicians willing to abide by the terms of the memorandum of understanding agreed to provide "ordinary medical care including examination; diagnosis and treatment; obstetrical care; emergency surgical care; and some hospitalization and ordinary drugs" to FSA families.[28] FSA clients often suffered from chronic disabilities, and while doctors were not required to do any remedial therapy, many did. Initially many families were unfamiliar with the insurance concept and hence called the doctor only in an emergency. In spite of concerns raised by participating physicians and medical societies about "abuse" (i.e., unrestricted demand for services) there was little evidence that this occurred. Alec Spencer, a rural general practitioner who participated in an FSA plan in Kentucky, recalled that "if you saw a patient then, you saw a sick person."[29]

The FSA's Health Services staff hoped that the medical care program would promote the concept of disease prevention among rural physicians. For ex-

ample, they expected that prorating bills would encourage physicians to provide more immunizations, nutritional advice, and prenatal and early infant
care. If these efforts succeeded, the argument went, families would be healthier, use fewer medical services, and leave more money available to reimburse
doctors. This rather naive view not only ignored the reservoir of untreated
chronic conditions the FSA knew existed among their clients, but it neglected the medical profession's characteristic emphasis on curative practice and
the longer time frame that preventive efforts often require to show financial
benefit.[30]

The FSA's policy was to allow county medical societies to set the fee schedule used in any given plan, but these were scaled down from the fees normally requested from low-income families. At the end of each month, bills for
services rendered were submitted either to a medical review committee representing the county medical society, local FSA representatives, and FSA families, or more simply to the trustee. If the submitted bills were in line with the
fee schedule, they were paid out of the allotted monthly fund. The committee was also responsible for resolving any conflicts that arose, such as differences between the doctors' bills and the fee schedule or apparent excesses in
service utilization or billings. Some county medical groups set up committees
to scrutinize the services and billing patterns of participating physicians, a
form of peer review that would become more prevalent later in the century.[31]

If the amount set aside for any given month covered all billings, doctors received 100 percent reimbursement. If, on the other hand, the submitted bills
exceeded the allotted sum, doctors received payment on a prorated basis.
Residual money at the end of the fiscal year was used to pay earlier prorated
bills. If no money remained, the unpaid balance was written off by physicians
as charity. FSA medical care leaders, in agreement with county medical societies, instituted the prorating feature of the medical care program as a form of
administrative control over excessive utilization and billing on the part of farmers and physicians.

Other administrative innovations originated from various local, state, and
regional field offices. In 1939, Steele Kennedy, an FSA regional Health
Services specialist in Arkansas, instituted the novel policy of automatically including membership dues as part of the annual rehabilitation loan. This policy caught the eye of agency administrators in Washington, since it immediately spurred the expansion of the medical cooperative program in Arkansas.
Kennedy was brought to the national FSA office in Washington, where he was
assigned the responsibility of coordinating the medical care cooperatives nationally. By 1940, the FSA was recommending to its field staff that annual dues
be routinely included in annual rehabilitation loans, applying nationally what
had started as a local policy. In his 1940 annual report, Chief Medical Officer
Williams urged field staff to allow exemptions to this policy "only when a fam-

ily shows cause why it should not take part in this plan which serves as protection not only to the family but to the government as well . . . Implicit in such a decision is the realization that the health program is an integral part of rehabilitation. Once that realization comes, the battle is half won."[32]

The inclusion of medical cooperative dues in rehabilitation loans benefited rural physicians as well as the FSA. The automatic membership in the medical care cooperatives served physicians' immediate financial needs by increasing the pool of money available to them. As a result, the dues policy generated little sustained opposition from rural doctors. From the FSA's perspective, an increase in membership improved the fiscal stability of the medical care program by spreading the economic burden of sickness over a larger number of families. Even with this policy in place, however, borrower participation in medical care cooperatives seldom approached 100 percent, and membership turnover was a common problem. In some counties membership was as low as 20 percent, and the FSA estimated that nationally only 50 percent of eligible borrowers participated in the medical care program.[33]

The FSA's policy of strongly recommending but not actually mandating the inclusion of membership dues in annual rehabilitation loans was significant. It is a good example of the FSA's continuing effort to improve service to its clients while keeping the support of local physicians and avoiding overt hostility from the AMA. All agreements between the FSA and county medical groups specified that participation by both physicians and clients was voluntary. Without this proviso, a confrontation with the AMA, whose vigilance regarding compulsory health insurance schemes was eagle eyed, would have been likely. By wording its new dues policy skillfully, the FSA was able to maintain the appearance of voluntary participation without giving up the advantages of strongly encouraging its clients to join the medical care plans.

The personnel of the FSA Health Services Branch took seriously their obligation to ensure that FSA families received competent medical care. For this reason, in part, that the agency decided to include only allopathic physicians in the medical care program. Midwives, chiropractors, and other local healers were routinely excluded from participation. There were rare instances when the FSA refused to sanction the continued participation of a physician in the plan because of recurrent complaints of incompetence—an unpleasant task that was left to district or regional field medical officers, rather than the local rehabilitation supervisors.[34]

Participating physicians provided medical care in their offices or in the patients' homes. Service was provided when requested by the patient, just as it had been before the cooperative was established. Parallel services that were part of the overall FSA rehabilitation program but not the medical cooperatives themselves included home visits from nurses and home management supervisors, and advice from sanitary engineers.

Miss Teal, nurse, bringing medicine for the treatment of hookworm to the Lewis family, rural rehabilitation clients. Coffee County, Ala., 1939. Marion Post Wolcott.

The agency produced many manuals and handbooks for use by its field staff. A *Group Health Services Manual*, published in 1941, was just one example of the FSA's skill in educating its staff and its clients. In step-by-step fashion, the manual outlined the services that a medical unit should include, such as "ordinary home and office care, . . . pre- and post-natal and obstetrical services, immunizations, such minor surgery as the practitioner is accustomed to performing, and all ordinary drugs usually furnished by him." Instructions were written in language designed to be easily understood and followed. The manual provided concrete suggestions on how to maintain good relations with local physicians. For example, supervisors were cautioned to "avoid becoming a nuisance . . . by annoying the physician with every little problem that arises," and to "remember the physician is only human . . . he enjoys an occasional pat on the back."[35]

The FSA worked with a class of farmers whose lives and livelihoods were traditionally controlled by forces beyond their reach—be they landlord, bank, or Nature. In attempting to break the cycle of dependence cultivated by years of marginal existence, often on submarginal land, the agency relied on a philosophy that emphasized personal and economic self-sufficiency. This was as true of the medical care programs as it was of the overall rehabilitation effort. For an agency whose ultimate goal was to help farmers assume greater control over their own lives, rehabilitation was an educational process and the health-promoting activities of the agency reflected this. *"Give the families a chance* to express themselves and ask questions when talking to them about the plans," the *Group Health Services Manual* instructed.[36] A 1938 memorandum to re-gional directors stated that "any plan in which the families unite to help them-selves should reduce the cost of medical aid and thereby make more effective the funds thus expended . . . a worthwhile beginning can be made by the *fam-ilies themselves* toward better health."[37] The FSA also asked its borrowers to sign a "health plan participation agreement" when they joined the medical care plan. The agreement set down in writing what health services were available, reinforced the importance of prevention, and showed the value the agency placed on the farmer's personal commitment to health as part of the rehabili-tation program:

> I agree to do my share in the Health Plan by following the advice of my doctor and dentist and of the state, District, and County Health Department, carrying out the food production and preservation program in my FSA Farm and Home Plan, keeping my home and outbuildings clean, free of insects and screened properly and keeping my well or other family water supply free of contamina-tion.[38]

The message that the FSA hoped to bring to its clients was a simple one: col-lective action could bring improved physical and economic health to FSA client families. FSA staff tried to encourage participation in the development and operation of the medical care cooperatives. As noted earlier, farmers pro-vided input into which services they would like to see covered by the medical care plan, and in some counties they sat on billing review committees. In ad-dition, plan members could bring grievances about physicians or services they received (or felt they should have received) to local FSA supervisors, or di-rectly to the local review committees for consideration. How successful this effort to involve clients was depends on whom you talk to. For T. G. Moore, it was an integral part of the program: "Even in those days, the days you're asking me about, consumers were involved. The people were involved in what hap-pened; they gave their input."[39] Other staff members, however, were frus-trated by their inability to involve FSA clients in the running of the medical care units. In 1943, Franz Goldman of Yale University conducted an inde-pendent evaluation of the FSA medical care programs in Arkansas, South Dakota, and Ohio, in which he expressed his disappointment that "participat-ing families had so little to do with the operation of the medical care plans." [40]

MEDICAL CARE IN FSA RESETTLEMENT COMMUNITIES

The most controversial of the FSA rural rehabilitation programs were the re-settlement communities, homestead projects, and greenbelt communities the agency built around the country. Begun by the Resettlement Administration, the program transferred individuals and families from land considered too poor to farm productively to land bought by the federal government and "sold" to the relocated families. In some cases, families were resettled because their farms were in the way of planned dam projects that would flood their land. Re-settled families were moved to federally planned, built, and funded model communities. These townships were quasi-autonomous in that they had elected community councils, but FSA managers exercised considerable in-fluence.

Resettlement was deemphasized when the FSA was created, but the con-struction of resettlement communities continued, albeit at a pace consider-ably below what Tugwell and his supporters had once hoped for. These mod-el communities formed only a tiny part of the FSA's overall rehabilitation effort, and many historians agree that they suffered from being "overplanned, underfunded, and poorly administered."[41] Nevertheless, the attention and notoriety they received was impressive. In one of the milder diatribes against the resettlement idea, a Memphis newspaper editor claimed that the FSA's continuation of Tugwell's utopian programs was proof that the FSA was "vac-cinated with the Tugwellian virus."[42] In spite of the criticism the program pro-voked, FSA administrators considered the resettlement communities unique social experiments that demonstrated the strengths and weaknesses inherent in the democratic process. In his 1938 annual report, Chief Medical Officer Williams wrote that the resettlement projects "constitute an example of con-sumer or social action to achieve a socially desirable goal—action which need not wait on governmental or professional initiative. They are voluntary, local, democratic, and strictly on a non-profit basis. They offer training value in the democratic process."[43]

The majority of the resettlement projects, subsistence homestead commu-nities, and greenbelt towns had medical care plans either on their own or in conjunction with a medical care cooperative for standard rehabilitation loan families. These medical plans shared many of the same features of the larger program, operating on a cooperative basis, with pooled membership dues and participation by community members on the governing boards of the cooper-ative associations. A number incorporated capitation. The range of services was decided by the community and negotiated with local physicians and med-ical societies. In at least forty resettlement projects, the FSA provided funds to build a community health center, and by 1940, some fifty-six resettlement communities had their own medical cooperative plans. These were usually

Sabine Farms, a Farm Security Administration project for Negroes. Dr. Lee's travelling clinic serving families. He is employed by the Sabine families on a cooperative basis. Vicinity of Marshall, Tex., April 1939. Russell Lee.

staffed by full-time salaried nurses and full-time or part-time salaried physicians. Occasionally, resettlement communities formed their own incorporated health associations and contracted directly with physicians, hospitals, and dentists for discounted medical services. Resettlement projects that were self-supporting, as some were, were able to employ a doctor (often fresh out of training) from a community fund generated from annual farm income. More commonly, the FSA provided salary support in order to attract and retain physicians. Some philanthropies (notably the Commonwealth Fund and the Kellogg Foundation) funded similar kinds of programs designed to bring physicians to rural areas and offered educational enhancement to those practicing there. Most of the physicians recruited were young and had been attracted to practice at least part time in the resettlement communities through advertising, guaranteed financial packages, and offers of free housing.[44] Agency statistics showed that salaried physicians saw more patients than doctors who were cooperating in medical care plans for rehabilitation borrowers, and this was one of the reasons why the FSA was willing to augment whatever support the community itself was able to provide. The apparent efficiency of salaried practice was also observed in the FSA migrant health plans and in the experimental health plan begun in 1942 in Taos County, New Mexico. The agency explained this phenomenon in part by noting that salaried staff provided more preventive services.[45]

UNDERCOVER INCENTIVES: PHYSICIANS' RESPONSES TO THE MEDICAL CARE COOPERATIVES

The FSA medical care cooperative program fulfilled many of the positive expectations of participating physicians and did not prove as bureaucratic as some had predicted. Rural doctors collected a substantial percentage of their billings from a group that had previously been able to pay very little—if anything—for medical care in the traditional fee-for-service system. Physicians' reimbursement varied from state to state and county to county, but averaged between 60 and 90 percent of billings according to statistics gathered by the FSA. Reimbursement for surgical services, hospitalization, and dental care tended to be greater on a percentage basis than that for general medical care. One especially diligent physician in Kansas maintained a detailed accounting of his practice in the years prior to the institution of an FSA plan in his county. It showed that he collected only 11 percent of his fees from farmers who later became FSA clients. After the FSA plan started, he received 61 percent of his fees from these same families.[46]

Judging from comments in medical journals, this new source of income was one of the key reasons that the medical community supported the FSA programs. One county medical society from Alabama claimed that "this class of patients pays more under this plan than when dealt with individually." Another practitioner threw his support behind the FSA plans because "it is a class of practice which I do and have done for many years and never gotten anything for; and now I get a little for what I do."[47] Lorin Kerr, a Public Health Service medical officer assigned to the FSA in the Pacific Northwest, later remembered how bad things got for some physicians: "There was a period in the thirties when there were doctors who were on soup lines," stated Kerr. "In the major cities there were doctors on relief! And in the South. So that there was an incentive, an undercover incentive, for the doctors to enter into that kind of agreement."[48]

As in the farmers mutual aid corporations, the benefits that doctors gained through their association with the FSA went beyond pocketbook issues, however, and touched directly on the allopathic medical profession's ascendancy over medical practice in the United States. The agency's policy of collaborating solely with allopathic physicians and working exclusively with county and state medical societies not only reinforced the economic condition of local

general practitioners, it also enhanced the social and political status of organized medicine. In 1942, for example, this was one of the reasons the Texas
State Medical Society cited for continuing its cooperative agreement with the
FSA in spite of considerable opposition from several county medical societies.
The FSA was "earnestly and sincerely cooperating" with physicians to address
any concerns, commented the *Texas State Medical Journal*.[49] A Kansas doctor
found the local medical care plan satisfactory, not only because it was financially advantageous, but because it also served the broader interests of organized medicine by putting control in the hands of the county medical society.
Addressing several disgruntled county medical groups in the neighboring state
of Oklahoma, he remarked that "we are a little ahead of them on the FSA. It
is $30 per year up in Kansas, instead of $15 a year per family; it is left entirely to the county societies."[50]

The FSA also gained support at the local level because many physicians,
and a number of state and county medical societies, were dissatisfied with the
apparent unwillingness of organized medicine's national leaders to address the
social, economic, and health consequences of the Depression. By the mid-
1930s, a schism in the ranks of organized medicine had clearly emerged. On
one hand were physicians who favored moderate reforms and a greater government role, at least for the medically indigent. On the other was the leadership of the AMA and fellow conservatives inside organized medicine who
steadfastly opposed reforms. This group was more likely to be composed of
wealthy urban specialists. Their argument for caution and restraint distanced
them from thousands of ordinary physicians, and many doctors simply felt that
changes in the financing of medical care were long overdue. In a number of
western states, including Washington, Oregon, and California, medical societies at the state and local level adopted relatively liberal positions on voluntary health insurance and government programs for the poor. Not coinciden-
tally, these were the same states where physicians' service bureaus were
particularly vigorous as well as where some of the more controversial private
health plans and consumer cooperatives took root, including the Kaiser-Permanente plans.

Dramatic and highly publicized examples of the rent in organized medicine
occurred in 1936 and 1937 when several prominent medical groups publicly
broke ranks with the AMA over health insurance. In 1936, a group of academic
physicians, supported briefly by the American College of Physicians and the
American College of Surgeons, formed the Committee of Physicians for the
Improvement of Medicine. In 1937, this influential group (sometimes referred to as the "Committee of 400") called for an expansion of the government's role in medical research, education, public health, and medical care financing for the indigent, prompting the *Nation* to quip that "Civil War has
broken out at last in the A.M.A.—a very civil and respectable war."[51]

Under Surgeon General Thomas Parran, the U.S. Public Health Service also moved beyond its traditional public health purview to involve itself directly in medical care issues and in the debate taking place on the government's role in medical care. Prior to Parran's tenure, the U.S. Public Health Service's relations with the AMA had been extremely cordial. Indeed, it was not uncommon for a retired Surgeon General to assume a top leadership position in the AMA. Now the agency's involvement in national health policy discussions, and as some saw it, Parran's bid to transform the agency into a national health department, placed increasing distance between it and organized medicine.[52]

The continuing activity of the Interdepartmental Committee to Coordinate Health and Welfare Activities influenced the attitudes of rural physicians and organized medicine toward the FSA programs as well. The 1936 election heightened Roosevelt's confidence in his domestic agenda, and soon thereafter he broadened the committee's mission to include an assessment of the nation's health needs, and subsequently to consideration of a national health program. In March 1937, the Interdepartmental Committee delegated these tasks to a small working group, the Technical Committee on Medical Care. The Technical Committee was chaired by Martha May Eliot, assistant chief of the Children's Bureau. Eliot and her colleagues set out "to study the whole range of the Federal relationship toward health and medical care activities in the U.S."[53] Although FSA Health Services Branch personnel were not on the Technical Committee, several members had close ties to the FSA.

The findings of the Technical Committee on Medical Care reaffirmed the earlier work of the Committee on the Costs of Medical Care and the Committee on Economic Security. The Technical Committee submitted its findings to Roosevelt in 1938 and they were later published in a monograph. This report called on the federal government to expand its role in the provision of public health services, including maternal and child health programs. The report also recommended increased federal dollars for hospital construction, development of a wage compensation system for workers with either permanent or temporary disabilities, and public funding for medical care for the indigent. The group's most controversial proposal was for the creation of a state-based insurance program, funded at least in part through public taxes, to provide medical care to the entire population. Although the Technical Committee on Medical Care did not propose a compulsory national health insurance plan, the group suggested that it be considered.

An enthusiastic Roosevelt responded to the Technical Committee's report by calling for a national meeting of all interested parties to consider its recommendations. In July 1938, a national health conference was held in Washington, D.C., with representatives from over 150 public, private, academic, and philanthropic groups, including personnel from the FSA Health Services Branch. Over the objections of the AMA, conferees endorsed most of the recommendations contained in the Technical Committee's report.

The pace and drama of these events electrified the medical community and forced the normally pugnacious AMA into a conciliatory mood. The AMA met quietly with the Interdepartmental Committee to Coordinate Health and Welfare Activities and agreed to abide by all of the recommendations if the proposal relating to health insurance was dropped. The Interdepartmental Committee, confident that the tide of public opinion had turned in favor of national health insurance, rebuffed the AMA's overture. In an effort to isolate compulsory insurance from other proposals that were more acceptable to organized medicine, the AMA called an emergency meeting of its house of delegates in November 1938 and rescinded its opposition to voluntary third-party health insurance.

In February 1939, Senator Robert Wagner of New York introduced a bill that would have implemented at least some of the Interdepartmental Committee's recommendations. By then, however, organized medicine's spirit of accommodation had passed. Buoyed by the conservative gains made in the 1938 election, the AMA attacked the bill in its entirety. Roosevelt was himself sobered by the election results, and he was already beginning to turn his attention to the European conflict. Much to the consternation of the bill's supporters, the president failed to use his considerable political power and personal charm to promote its passage. Extensive and contentious committee hearings were held, but the legislation died without reaching the floor of Congress.

The gulf between the AMA and practicing rural physicians was central to the FSA's ability to build its medical care programs. The comments of frustrated and disappointed physicians filled the pages of medical journals, many of them castigating organized medicine for abdicating responsibility in the face of a national crisis. The exasperation of an Alabama general practitioner in a 1939 article in the *Journal of the Medical Association of the State of Alabama* is typical of the tone and intensity of the sentiments of a large number of rural physicians: "These men hold up their hands in holy horror and decry social or state medicine when plans of this sort are mentioned," he wrote, "but never submit any plan which would lift the profession out of the economic rut in which it has found itself."[54]

Overburdened rural physicians genuinely appreciated the FSA's advocacy on behalf of poor rural families and those rural practitioners struggling to meet the health needs of their communities. At the annual meeting of the AMA house of delegates in 1942, a Kentucky physician rebuked the FSA's critics, claiming that "the FSA is making a notable contribution . . . We will welcome with open arms in Kentucky the FSA when it comes."[55] In 1939, an Alabama physician defended his county medical society's participation in FSA programs on both humanitarian and economic grounds:

> Years of crop failures, crop reductions, debt and one crop ideas have impoverished them to such an extent that a majority of the tenant farmers can no longer make enough to buy the necessities of life, much less pay for medical care . . . the physicians of Wilcox County welcomed any method that would extend to these families medical care and to the physician, aid in performing his duties.[56]

DISCOURAGING SITUATIONS: EARLY DIFFICULTIES IN THE MEDICAL COOPERATIVES

Through a combination of tact, enthusiasm, and economic coercion, the FSA expanded the medical care cooperative program up to 1942, when it peaked at 1,200 medical plans in 41 states, covering more than 625,000 individuals. The convergence of farmers, physicians, and the FSA's self-interest proved decisive in many areas, but not all. During the seven years of sustained growth from 1935 to 1942, the FSA confronted wariness—and sometimes blunt opposition—from individual physicians and their organized representatives at the state and local levels. Closer examination of instances where medical care plans experienced fitful starts, anemic growth, tepid support, or out-and-out rejection presents a more complex view of the successes of the medical care program. More important, it locates the roots of the agency's future difficulties in conflicts that existed well before World War II became a significant factor.

State medical societies in Illinois, Indiana, Massachusetts, Pennsylvania, and New York prohibited county medical groups from negotiating with the FSA until they had taken a position. In several of these states, the journals that typically carried the transactions of the state medical society give no indication that any formal decision was ever made one way or the other, and this fact was reflected in the relative paucity of FSA cooperatives in New England and the eastern seaboard states. In Indiana, Ohio, and Illinois, only 116 county medical societies had signed memoranda of agreement with the FSA by 1941, although a year earlier the FSA's chief medical officer had predicted that 300 county medical care units would soon be operating in these states.[57]

In Pennsylvania, there was a slight wrinkle to this pattern. The FSA approached the state medical society in 1937, but the group declined to make a decision and specifically forbade the FSA from discussing its proposals with county medical groups. In 1938, however, the Pennsylvania Medical Society looked the other way when the FSA worked out an arrangement with the Crawford County medical society. A year later, the state medical journal took note of the favorable reimbursement rate for Crawford County physicians, and the FSA pried an agreement out of the reluctant state medical society, provided there was "constant emphasis" on the emergency nature of the program and county medical societies approved each medical care plan.[58]

The reasons that county or state medical societies refused to cooperate with the FSA were sometimes noted and in other cases may be inferred. Economic themes predominated. Some county and state medical societies hesitated because they suspected that prorating and unlimited demand for services by FSA clients would drive reimbursement rates down to unacceptable levels and trap doctors in a situation where they were required to give services for diminishing payments. The comments of one participating medical society that the "arrangement is specifically stated to be an 'understanding' and 'not a contract' between the corporation and the association" speaks to this concern.[59] The FSA was sensitive to this issue and judiciously called their agreement with each medical society a "memorandum of understanding."

There were instances in which medical societies' bylaws precluded the organization from entering into any contract on behalf of chapter members. This was the case in Augusta County, Virginia, where the medical society was unable to sign a memorandum of understanding, but did endorse the medical care cooperative and allowed its members to sign individual agreements with the FSA.

One of the greatest obstacles the FSA faced was the unease many physicians felt about third-party insurance. A 1938 letter written by Assistant Chief Medical Officer Fred Mott to Ralph Williams captured this anxiety: "Dr. Fisher appeared to be more in sympathy with our aims than Dr. Robertson, but even he seemed disturbed by the prospect of having physicians, in effect, contract to look after families for a definite, limited sum of money. Both men stated their opposition to any pooling of funds, although they gave no convincing argument for their stand."[60] In another instance, a regional director wrote to Will Alexander that "numerous contacts with the medical associations in the seven states in this region, and the officers of the AMA" led him to conclude "that it was impossible to get the cooperation of the profession in the setting up of these formal associations. They are willing, however, to provide for our clients medical care based on the client's ability to pay. An agreement is to be worked out on an individual basis between the physician and the client."[61] This was not at all what the FSA had in mind. FSA leaders preferred to forgo any medical care effort rather than accede to the wishes of those medical groups unwilling or unable to get beyond the traditional sliding-scale fee-for-service medical care.

In some cases doctors who supported the program denied that it used insurance, preferring to see it as sort of a gentlemen's agreement between the medical profession and the government. "They wanted us to make an insurance problem out of it, and it is not an insurance problem," commented a Texas physician in 1942, "it is a partnership, a simple partnership for the purpose of taking money, however we can get it honestly and honorably now and dividing it among ourselves for an honest and honorable service."[62]

The architects of the FSA medical care programs at the time and in later reminiscences acknowledged that the novelty of health insurance as a means of more evenly distributing the costs of medical care made it more difficult to promote the program. A harsher assessment was later offered by one FSA staffer, who felt that "the professionals in those days who had been brought up in this free enterprise system could never learn or appreciate what was needed in a coordinated social health program . . . It was beyond them."[63]

When faced with unrelenting opposition from state or county medical groups, or from individual physicians, FSA supervisors and the agency's Health Services staff had few options. In some cases, the FSA simply waited and hoped that the experience of other medical societies would encourage a more compliant attitude. In other instances, the agency took an indirect route in an effort to move the process forward. In some communities, the FSA approached wary county medical groups with a simple request to allow local physicians to provide a one-time medical examination to prospective FSA clients in order to determine medical fitness before their loan was approved. Not infrequently this very limited experience with the FSA allayed professional concerns sufficiently to enable the more comprehensive plan to be organized.

Even when physicians' support of a medical care plan was obtained, this approval was often neither unanimous nor irrevocable. At the local level, and very early on, doctors' attitudes were fluid and readily influenced by changing financial, social, and political circumstances. Letters hinting at the difficulties that their personnel were encountering at the local level began to accumulate in FSA regional offices as early as 1938.

In Boone County, North Carolina, doctors postponed the start of an FSA plan for ten months in 1938 while they debated the proposal. By the end of December, the county medical society approved the FSA plan, but nearly a year later, the Boone County cooperative had yet to enroll any FSA clients. Then in the fall of 1939, the Boone County medical society reversed its earlier decision and revoked the memorandum of understanding. Finally, after further negotiations the agreement was reinstated, and by 1940 Boone County had a medical care unit in place.[64]

In the 1938 *Annual Report of the Chief Medical Officer,* Ralph Williams noted that FSA personnel were blocked from starting a medical care plan in some communities or experienced difficulty maintaining them. In later annual reports, corroborated by extensive intra-agency correspondence, physicians' opposition and plan terminations featured more prominently. In 1941 alone, forty-four medical care units were terminated, representing 8.1 percent of the FSA medical care plans operating that year. Based on available but somewhat incomplete statistics, plan terminations averaged 10 percent per year after 1941. Those regions where the medical care programs' presence was greatest had even higher termination rates. Region IV, covering the southern states of North Carolina, Tennessee, Kentucky, South Carolina, Georgia, and Alabama, experienced an annual termination rate of over 13 percent from 1939 on. In North Carolina, termination averaged almost 18 percent over the same period.[65]

Medical care units that were successfully initiated could also be derailed in short order by physicians' dissatisfaction. Often the crux of doctors' unhappiness was economic: specifically, low fee schedules and the prorating of bills. Intra-agency correspondence and the annual reports of the chief medical officer indicated that reimbursement issues were the primary cause for physicians' opposition and plan termination. In North Carolina, the FSA's regional director noted in a 1939 letter that the program was generally doing well, but that problems emerging around economic issues were threatening to disrupt it.[66] When Marion Goff, a county rehabilitation supervisor, visited Madison County, North Carolina, in 1939, he "found the situation there very discouraging." Local physicians were threatening to rescind the memorandum of understanding because "insufficient funds had been advanced." In a letter to the regional director, Goff argued that additional funding was unwarranted. "Since the doctors do not care to proceed," he wrote, "there seems to be nothing that we can do."[67] That same year, another North Carolina county medical society passed a resolution refusing to cooperate any further with the FSA, and rebuffed the FSA regional director, who had offered to meet with them to discuss their decision.[68]

Commentary on the FSA medical care program in state medical journals in the period from 1940 to 1942 supports the view that economic concerns were largely responsible for physicians' dissatisfaction. According to a comment in the 1940 *Journal of the American Medical Association,* a group of Louisiana physicians revoked their memorandum of understanding with the FSA when prorated reimbursements declined too far: "Since they could not make a living collecting 36 percent of their fees, they did not therefore feel justified in continuing the FSA plan. As a whole the patients did not abuse the service with excessive demands and the doctors tried to keep unnecessary visits to a minimum . . . Since it was believed this group could not contribute any larger amounts, the doctors decided that they preferred to charge and collect from each patient as an individual."[69]

In 1942, the Medical Association of the State of Alabama refused to renew its agreement with the FSA, citing increasing service demands and declining reimbursement. In some instances, physicians attributed low reimbursement rates to lack of discipline on the part of participating physicians—specifically, billing excessively and providing unnecessary services. Often, however, they focused blame for these two problems on the insurance mechanism which, they argued, encouraged utilization by reducing out-of-pocket costs to patients and limiting the amount of bad debt likely to be incurred by doctors for the care they provided.[70]

Less commonly, the FSA withdrew support for a medical care plan, usually when the memorandum of understanding was being routinely violated. The 1940 *Annual Report of the Chief Medical Officer* reported that "lack of satisfactory internal control" (an FSA euphemism for a situation where physicians were charging fees above the set schedule) was responsible for the FSA's decision to terminate four programs in Iowa and Indiana that year. In other counties, the chief medical officer complained that the billing review committee was indiscriminately approving "any and all bills submitted" and forcing the FSA to withdraw its support. Overcharging by physicians provoked FSA personnel to terminate an agreement with an Oklahoma county society in 1940.[71] In 1941, when physicians in three separate plans in Tennessee and North Carolina demanded 100 percent payment (even though they were receiving 80 percent of total billings to that point) the FSA terminated all three plans.[72]

After 1940, the exchanges between the FSA and local and state medical societies about financial issues grew more heated, and disagreements about the medical care program divided many state and county medical groups. This was reflected in a number of medical journals where commentary on the FSA plans from county medical societies ranged from hostility to ambivalence to grudging support.

Claiming widespread dissatisfaction among its membership, the state medical society in Alabama abrogated its agreement with the FSA in 1940. At the same time, the organization still allowed county medical societies to participate if that was their preference. One Alabama county medical society was riven by internal controversy on the matter. According to the secretary of the state medical society, one county society viewed the program as "a total cheat" and had accused the FSA of "chiseling" local doctors. Whether he intended to be amusing or not, the state secretary added that while the FSA plan "caused considerable arguments in the society . . . to date there are no casualties" and that he and his colleagues had decided to "continue it for another year, maybe."[73] Similar fractures occurred between county and state medical societies in other states, including Kentucky, Iowa, and Arkansas.[74]

In Texas, a state with a large FSA presence, half of the county medical associations approached by the FSA refused to sign agreements, and a Texas doctor told his colleagues at a 1940 AMA meeting that physicians in Texas were not "sold, hook line and sinker, on any FSA plan."[75] A survey undertaken by the Texas State Medical Society in 1942 demonstrated the ambivalence with which the medical community viewed the programs. When asked if the

FSA was "more interested in better medical care for clients with adequate compensation to the physician or in securing ordinary care for the client with less compensation to the physician," a large majority responded that the plans provided better care *and* fair reimbursement. But when county groups were asked whether they would like to "curtail, expand, or abolish" the FSA plans, a majority preferred to end the FSA experiment. In the face of this contradiction, the Texas State Medical Society continued collaborating with the FSA.[76]

Rhetoric claiming that the FSA plans violated "medical ethics," "provided inferior care," and that "medical service does not seem to lend itself to cooperative handling" was sprinkled throughout numerous state and national medical journals.[77] Beneath this rhetoric was physicians' discomfort with a program that operated contrary to norms, namely, private fee-for-service practice. Doctors were not the only ones struggling to adjust to the FSA's use of health insurance. FSA clients who refused to join the program or who let their memberships lapse often did so because they did not want to pay for medical care without assurances that they would get their dollars' worth of service. FSA medical leaders sometimes faulted Rural Rehabilitation supervisors for not understanding the goals of the medical care program, or for not adequately explaining the purpose of the medical care plans to FSA clients.[78]

Further complicating physicians' concern about introducing the insurance principle into medicine were the recurring attempts by some in Washington to institute compulsory health insurance. For some doctors, the FSA programs, whatever their bad points, were a more acceptable alternative. The continuing inaction of the organized medical profession and its leaders in dealing with the medical problems of the poor risked their losing ground to those favoring solutions fraught with more dangerous potential. As one rural Tennessee physician warned his colleagues at the 1938 annual meeting of the American Medical Association: "There is going to be more socialization of medicine, and unless the AMA and the Tennessee State Medical Association plan to meet the needs of the group who possibly cannot secure medical care . . . a system of state medicine will be crammed down our throats."[79]

Other physicians and organized medical groups were less certain of the FSA's benign intent. These skeptics saw the FSA as part of the New Deal's goal of instituting a national health program, not as an alternative to it. Historians have pointed to Roosevelt's failure to give his full support to national health insurance as an indication that he did not really favor it.[80] Physicians at the time, however, believed that instituting such a program was one of the president's goals. Doctors needed to look no further than the Committee on Economic Security or the recommendations coming out of the Interdepartmental Committee to Coordinate Health and Welfare Activities to be convinced of what *California and Western Medicine* called the New Deal's "desire to control everything, even to the death rattle of humanity."[81] At the annual meeting of state medical society secretaries in 1940, a physician told his colleagues that he, for one, did not believe that FSA plans were as temporary as many of his colleagues hoped:

> It is fair to say that the program of the FSA in dealing with the rural population groups is not an experimental thought . . . I was impressed in all the contacts we have had with the FSA that a great many of its clients are in no different financial circumstances than the vast majority of the people living in the same county . . . once we enter on a plan to accept this principle I have no doubt at all that the demand . . . would be extended to all other people in similar circumstances . . . It seems to me that the plans of the FSA need to have a real consideration when we discuss a subject as broad as rural service.[82]

Doubts about whether the FSA programs were only an emergency insurance plan for the rural poor antedated the intensification of the debate over national health insurance during World War II. The 1940 comments of a participating Kansas doctor suggest that for some physicians this concern was quite real. "If we aren't careful," he cautioned, "this FSA program will set the pattern and will have decided the future of medicine."[83]

GOING UP THE LADDER OF SUCCESS:
THE IMPACT OF THE MEDICAL CARE COOPERATIVES

While the impact of the FSA rehabilitation programs on rural poverty is disputed, even the most critical scholars acknowledge that they provided immediate relief and in many cases long-term benefits to participating families. According to historian Theodore Saloutos, "Farm families made rapid gains in their net worth, standards of living, and abilities to support themselves" as a result of the FSA rehabilitation effort.[84] This view is buttressed by a 1939 survey of nearly 360,000 FSA rural rehabilitation borrowers in which it was found that their net worth exceeded their net debt by $83 million. In addition, somewhere between 80 and 90 percent of the loans made by the agency were repaid in full, a stunning achievement in light of the extreme poverty of most borrowers.[85]

Assessment of the impact of the medical care programs on the health of FSA families is hard to quantify. For reasons summarized in the preceding discussion, nearly two-thirds of all FSA rehabilitation borrowers could not join a medical care plan because they were never organized in their community. Even where plans existed, a significant number of borrowers chose not to participate or participated only intermittently. Failure to understand the insurance principle was responsible for many families' decisions not to join or to drop their memberships, but other factors played a part as well. In some cases the doctor a family was used to dealing with was not part of the program or lived across county lines. Other families preferred the care of osteopaths, chiropractors, or midwives, who were specifically excluded from FSA plans. For the families who did participate in the medical care programs, however, statistics gathered by the FSA in conjunction with the U.S. Public Health Service provide some support for the positive impact of the programs. Much of this information reflects the FSA's emphasis on promoting health and preventing illness.

A series of articles written by statisticians from the U.S. Public Health Service between 1938 and 1942 showed that for some conditions there were demonstrable improvements in the health of FSA families. For example, the prevalence of two diet-dependent conditions, iron deficiency anemia and pellagra, decreased in FSA families.[86] An extensive health survey in southeastern Missouri found that hemoglobin levels in 843 rehabilitation families improved, with greater gains experienced by African American FSA clients than by white borrowers. Hemoglobin levels in black families, the article commented, "improved steadily as the length of time spent in the FSA programme [sic] increased."[87] Historians have also credited the FSA's nutritional and health education programs with improving rural dietary habits, thus contributing to the decline of pellagra in the South.[88]

Much of the evidence, however, is anecdotal. "It is not possible to state the specific positive contribution made by the health associations to the rehabilitation of members," wrote Chief Medical Officer Ralph Williams in 1941, "but it seems a fair judgment that without such health services a considerable number of the rehabilitation borrowers would be in a worse position economically and physically than they are now."[89] Letters the agency received from field staff and borrowers spoke to the value of the health care programs. One FSA publication quoted an FSA borrower who recalled that he could not afford to call in a doctor when one of his sons contracted tetanus, or "lockjaw." Once the farmer joined the local medical care cooperative, he and his family enjoyed a measure of security that made a world of difference: "If we'd of had the doctor in time, maybe our boy could've been saved. Now it's different. My wife here has been laid up with heart troubles and if we didn't have a medical plan she couldn't get to the doctor. We had a bad year. The fruit didn't bring much and the tomatoes were spoiled by the weather. You never know what you might run up against."[90] Similar testimony can be found for the resettlement communities. At the Crew Lake Homestead medical care cooperative, a physician reported that "the families require less care now than they did 2 years ago, that they abuse their privileges very little, that they call him early in illness, that he enjoys making repeat calls without feeling it a burden on the families, and that the general Farm Security Administration program has already shown results in healthier families, particularly the children."[91] A group of women from the predominantly black Haywood Farms resettlement project in Stanton, Tennessee, sent a letter to their home management supervisor thanking her for "the protection of their health" and for taking them "up the ladder of success."[92]

The FSA's primary objective in developing a medical care cooperative program—to ensure that its clients had access to medical care and to protect government loans—was largely achieved. The agency was able to use physicians' combined financial, humanitarian, and political self-interest to great advantage in promoting its medical care programs from 1935 to 1942. The FSA also managed to stretch traditional norms of medical practice and financing in rural areas.

Over time, FSA leaders extended their influence over the management, financing, and organization of medical care in rural communities. For example, the agency developed an extensive medical care delivery program in agricultural communities dependent on migrant labor, and an experimental health plan that was open to both FSA and non-FSA families. In both of these programs, the FSA assumed much greater administrative control than it had in the medical care cooperative program, a development that pushed the limits of the carefully negotiated relationship between doctors and the government. This, in turn, gave rise to more visible conflict between physicians and the FSA at local, state, and national levels.

CHAPTER THREE

A Long and Far-Reaching Program

The FSA Migrant Health Plans

I worked in your orchards of peaches and prunes,
Slept on the ground in the light of your moon.
On the edge of your city you've seen us and then,
We come with the dust and we go with the wind.
WOODY GUTHRIE, "PASTURES OF PLENTY"

John Steinbeck's fictional account of the Joad family in *The Grapes of Wrath* provoked outrage and controversy when it was published in 1939, and helped thrust America's migrant workers into the national limelight. Migrant farm workers were the most destitute of all agricultural groups during the Depression. Like the Joads, hundreds of thousands of rural families left their farms, loaded their meager belongings into jalopies, and set off in search of work. Most of them headed west to California.[1]

The majority of the migrants lived in makeshift camps by the side of the road, or in substandard housing provided by local growers. Some found shelter in the temporary tent camps erected by state and federal governments. In 1935, however, the California Emergency Relief Administration built two permanent farm labor camps at Marysville and Arvin using federal and state relief funds. The decision to build permanent structures represented an important departure for the New Deal's relief programs for migratory farm workers. Although stories about the early days of the federal farm labor shelter program abound, few are as colorful as that provided by Lorin Kerr, an FSA field medical officer who worked with migrant health programs in the Pacific Northwest. As Kerr tells the tale, the idea of building permanent camps came from a camp manager who finally decided:

> By God what they had in the way of tents was just no good. So what he did was contract with somebody on the local business level to build a camp out of wood. To have places where people could live in homes, where there would be meeting halls, buildings for privies or toilets and showers, where people could do washing and have their meals together if they wanted. The whole schmear. When it was all done, it cost $50,000. So he sent the bill back to Washington and he said "I recommend that this be paid." Well, shit, no one had ever done

The new migratory camps now being built by the Resettlement Administration will remove people from unsatisfactory living conditions such as these and substitute at least the minimum of comfort and sanitation. Vicinity of Marysville, Calif., April 1935. Dorothea Lange.

anything like this before. So [Assistant to the Secretary of Agriculture] Beanie Baldwin screamed and went off and looked at it and came back [to Washington]. They pushed a bill through Congress to get that $50,000 paid right away, they thought it was so good. And that was the beginning of the migratory camp program.[2]

Rex Tugwell was lukewarm to the concept of farm labor camps when the Resettlement Administration first took control of the Federal Emergency Relief Administration's rural rehabilitation programs in April 1935. To Tugwell the camps were an indirect subsidy to private growers and encouraged them to decrease their already meager support for the men, women, and children who harvested their crops. Farm labor camps provided minimally acceptable standards of housing and sanitation, but they did little to promote land ownership, which Tugwell saw as one of the RA's primary missions. Pragmatists in the RA countered this view by arguing that the camps were a natural adaptation of the government's overall rural rehabilitation agenda and that rehabilitation in areas dependent on farm labor would necessarily assume a different form from that followed elsewhere.[3]

After a wrenching tour of California in 1936, Tugwell reversed his position. He was appalled at the living conditions in most private growers' camps and the rank squalor of the hundreds of roadside camps and squatter villages that were everywhere in evidence. These contrasted powerfully with the two federal camps. As Tom Joad said after the family arrived in the federal camp at Weedpatch: "Ma's gonna like this place. She ain't been treated decent in a long time."[4] Soon after Tugwell's visit, niggardly appropriations for camp construction were rapidly increased. The Farm Security Administration expanded the farm labor camp program after 1937. By 1940, there were forty-two federal camps housing approximately 300,000 migrant families, most of them in California, Arizona, and Texas. However, it was the severe labor shortages caused by World War II and the importation of farm labor from neighboring countries that provided the biggest boost to the farm labor shelter program. By war's end, there were some 250 permanent and temporary federal camps scattered across the country and the migrant health program operated in some fashion in nearly every state in the union.[5]

Adjacent to the permanent camps, the government carved out small farms of 80 to 300 acres where long-term camp dwellers could rent homes and farm collectively. This modification of the original shelter concept represented a graded increase in responsibility and permanency. This was more to the liking of those who believed that the government's aim should be to help low-income farm groups achieve social and economic self-sufficiency. According to FSA camp manager Henry Daniels, his real job was "to move these people out of the migrant stream and resettle them so that they would work in the general area and be able to take part in the community rather than keep moving."[6]

Large growers saw the increasing federal presence in their communities as a threat to their traditional dominance and at times their resentment flared into violent opposition. T. G. Moore, who continued his involvement with the New Deal rural rehabilitation programs in the farm labor shelter program in Texas and Colorado, had personal experience with the antipathy the federal program elicited from the agricultural establishment. Before the federal camps opened, "there had been some charges of peonage assessed against some of the big growers," recalls Moore, "When we came in they were no longer able to control the workers the way they felt they should be able to."[7] Although Moore disputes it, there are those who claim he was Steinbeck's inspiration for the federal camp manager immortalized in *The Grapes of Wrath.* Lorin Kerr recounted an episode in which Moore stood down a local sheriff intent on breaking a strike by Mexican workers. Moore recalled the experience in less dramatic terms, but acknowledged that he was eventually "booted out of the state" for getting into it "crosswise with the big sugar beet companies." Henry Daniels, a camp manager in Texas, later remembered he had to "get out from under the gun" literally as well as figuratively when he aroused the opposition of powerful local growers.[8]

The influence of the federal camp program on overall living conditions for migrant workers has been argued since its inception. Certainly the federal migrant camps were too few in number to support more than a small fraction of those migrants in need. Even at their wartime peak, federal camps were dwarfed in number and size by employer-owned camps. Still, those fortunate enough to find shelter in the federal camps were noticeably better off than those outside the camps. In 1942, Carey McWilliams wrote that while the FSA program was "wholly inadequate," it had kept "thousands of migrants alive since 1938."[9]

The impact of the farm labor shelter program extended beyond the boundaries of the camps and thus surpassed what absolute numbers alone would suggest. For example, the presence of the government camps was an incentive to private growers to improve housing and living conditions for their workers, especially during World War II. The government brought its substantial weight to bear on important regional labor issues as well, supporting unionizing efforts and the minimum wage. More intangible benefits, such as the opportunity to move from a transient lifestyle to landed farming, the involvement of migrants in camp governance, and broad educational programs that the FSA incorporated into the life of the camp community were important achievements as well.[10]

In 1938, the FSA bolstered its commitment to making the health of migrants central to its overall rehabilitation effort. Following the model used in the medical care cooperative program, migrant health plans initially started with individual arrangements between the FSA and county medical societies. However, the medical care program for migrant workers evolved into a more coordinated national effort in 1938 when the first of the quasi-independent migrant health plans, the Agricultural Workers Health and Medical Association (AWHMA), was created in California and Arizona. Over the next eight years the FSA's migrant health program expanded nationally, including formal regional or statewide agreements creating a total of seven agricultural workers health associations, or AWHAs.

The AWHAs provided comprehensive medical services, including hospitalization, nursing, and physician care. Through them the FSA's involvement in public health also expanded significantly. All FSA medical care programs were multidisciplinary. They incorporated nutrition, health education, and sanitation into traditional allopathic medical care. It was in the migrant health programs, however, that this approach was most fully realized. Both doctors and FSA camp nurses provided acute medical services, and nurses also functioned as health educators and sanitarians. The expertise of sanitary engineers and home management supervisors was also routinely utilized. Another well-developed feature of the migrant health program was its hospital coverage. The AWHAs used a number of novel administrative features that became far more common later in the century. These included preauthorization for hospitalization, surgery, and specialist consultations, and limitations on hospital length of stay.[11]

CALIFORNIA FIRST: THE AGRICULTURAL WORKERS HEALTH AND MEDICAL ASSOCIATION

The largest and most fully developed AWHA was the Agricultural Workers Health and Medical Association in California and Arizona and its archival records are especially rich. Its history is significant because of the powerful ally the FSA found in the California Medical Association (CMA). For a number of years, the CMA adopted more lenient positions on a number of health-related reforms, including voluntary health insurance, than did the American Medical Association. The CMA's close working relationship with the FSA enhanced the somewhat rogue reputation that California physicians acquired in this period.

In the summer of 1938, FSA Chief Medical Officer Ralph Williams made an investigative trip through the San Joaquin and Imperial valleys of California to assess the health status and medical care in agricultural communities with large migrant populations. By this time, there were thirty federal farm labor camps in California (and six in Arizona). Williams visited many migrant camps—private, federal, and roadside—during his trip. He met with state public health and welfare officials, as well as private and federal relief officials, to determine their views on how the FSA might provide medical relief to migrant workers in these camps. Williams also met with numerous local public health officials, welfare agencies, and physicians who were struggling under the mounting tide of in-migration to the region.[12]

Williams was no stranger to rural poverty. He had done sanitation work for the U.S. Public Health Service in the rural South before he had come to Washington. Nevertheless, the conditions he witnessed stunned him. With the blessing of Surgeon General Parran, he immediately assigned eighteen nutritionists and six public health nurses from the Public Health Service to work in the migrant camps. Williams also authorized the use of medical social workers to coordinate camp registration and eligibility, and dispatched several Public Health Service medical officers to help the California Department of Public Health coordinate medical services to the state's burgeoning migrant population.[13] Drawing on the agency's experience in the Dakotas, Williams floated the concept of a statewide incorporated health association to those with whom he met and generally found a favorable response to the idea.

Fortunately both CMA President Howard Morrow and Karl Schaupp, chair of the influential Committee on Medical Economics, were willing to cooperate in the matter. As in most medical societies, the Committee on Medical Economics determined the organization's position on matters of health care reform, financing, or legislation, so Schaupp's support was vital to the success of the FSA's medical care program. Schaupp later became president of the CMA, and he remained a steady friend of the FSA migrant health program.

Throughout the fall of 1938, the FSA and the CMA grappled over fee schedules, scope of services, and other issues of professional control before reaching final agreement in late November. By midwinter, the Agricultural Workers Health and Medical Association was incorporated with a federal grant of $100,000. The agreement created a seven-member board of directors, composed of physician and federal representatives. Medical issues were left in the hands of the CMA, while fiscal matters were arbitrated by the FSA. The California Medical Association created its own physicians' advisory panel, chaired by Schaupp, to oversee the development of the initiative and report regularly to the CMA's executive council.[14]

At the insistence of the CMA, the Agricultural Workers Health and Medical Association initially provided reimbursement for medical services directly to the migrants, who then paid their physicians. This was a variation on fee-for-service not unlike the grant system initially utilized by the Federal Emergency Relief Administration. However, the mobility of migrants soon caused delays in processing medical care grants, slowed physicians' reimbursement, and in turn generated resentment by participating doctors. After further negotiation, the original agreement was modified and from that point on, providers of health services—be they physicians, hospitals, or dentists—submitted their bills to the Agricultural Workers Health and Medical Association and received payment directly from the association.[15] In short, the association became a federally funded third-party health insurer. Although California doctors were apprehensive about governmental incursions into medicine, the use of an incorporated organizational, administrative, and fiscal entity preserved the appearance of independence and thereby eased their fears.

Over the next six years, similar models evolved in five other regions with heavy concentrations of migrant labor. In 1941, an agricultural workers health association gave migrant workers in the states of Washington, Idaho, and Oregon access to free medical care in federal camps; in 1943 the Texas Farm Laborers Health Association, the Migrant Labor Health Association of Florida, and the Atlantic Seaboard Agricultural Workers Health Association were established. The last of the AWHAs, the Midwest Agricultural Workers Health Association, which covered migrants in Ohio, Indiana, and Illinois, was created in 1944. A Great Plains AWHA that covered North Dakota and South

Dakota existed at some point, but the dearth of archival references to it suggests that it must have been very short lived.[16]

Accurate estimates of the number of individuals served by the FSA migrant health plans are difficult to come by. The agency's own figures suggest that between 75,000 and 200,000 migrants were covered by the plans at any given time. Because of the migrants' mobility, normal seasonal fluctuations in employment, and the ability of any migrant to use the medical facilities for treatment, the total number of doctors' visits over the course of any given year certainly exceeded this figure, probably severalfold. In Texas, for example, the 3,800 U.S. citizens and their dependents covered by the Texas Farm Laborers Health Association in 1946 had a total of 13,000 "contacts" with the association. [17] During the war years, the percentage of citizens covered by the migrant health plans dropped sharply, and imported farm laborers comprised approximately 60 percent of the total number of eligible farm workers.

CLINICS ON WHEELS: MEDICAL CARE IN THE AGRICULTURAL WORKERS HEALTH ASSOCIATIONS

In setting up the migrant health programs, the FSA drew from its experience with medical care cooperatives. Still, there were distinctions that set the agency's two largest medical care programs apart from each other. For example, migrants did not bear any direct costs for the services they received in the AWHA, whereas rehabilitation clients in the medical care cooperatives paid annual dues. The FSA negotiated a memorandum of understanding with the state medical society for each agricultural workers health association and the agency still needed the approval of county medical groups in the counties where a migrant clinic was opened. FSA camp managers, field or district medical officers, and often Health Services specialists from Washington or regional offices, worked out agreements with individual county and state medical societies in creating an AWHA. Local physicians were then recruited by camp managers and agreed to render services using the discounted fee schedule (negotiated with the county or state medical society) and submitted their bills to the AWHA for payment. Professional participation was voluntary and by and large cooperation was the rule. Migrants were free to choose their doctors, although in practice there was little choice. Nurses sent patients to participating physicians' offices at their discretion or migrants simply saw the doctor staffing the camp clinic on any given day.[18]

The FSA built clinics in the permanent migrant camps run by the government, each with waiting and examination rooms, and a small dispensary. The Agricultural Workers Health and Medical Association was distinct from the other AWHAs in having association-owned and -operated medical clinics close to but not always in the camps themselves. Camp health clinics varied in their size, their level of staffing, and the hours during which they operated. In a number of western states, the FSA accommodated the mobility of migrant workers by operating moveable health clinics, dubbed "clinics on wheels."[19] Trailer clinics were hitched to trucks and transported from crop to crop along with the migrants themselves. Mobile farm labor camps were used as an adjunct to the permanent camps in areas where shorter harvest seasons made it less reasonable to erect permanent camp structures. The mobile camps typically held 200 tent platforms, all moveable and transportable; a first-aid clinic built in a mobile trailer, and portable sanitary facilities, including showers and baths. Permanent camps were more feasible in communities with longer or serial harvest seasons for the same crop, or where different crops required harvesting within the same geographic area.[20] By January 1942, medical clinics were accessible to the families and workers living in California's thirty-five standard (i.e., permanent) farm labor camps and twenty-three mobile shelter camps. These health facilities were accessible to migrants living outside the federal camps as well.

While there were important local variations, the overarching organization and philosophy of the migrant health programs were consistent from region to region: comprehensive acute medical care, government subsidization of costs, emphasis on prevention and health education, and accommodation of the local medical community. Coverage for health conditions was generously defined. Health services provided to agricultural workers were comprehensive; they included medical, surgical, hospital, and dental care; prescription drugs; diets; and nursing.[21] Preventive services and public health measures were also provided by AWHA staff or coordinated through the program. These included immunizations, prenatal and postnatal care, health education, and even sanitation services.

> Care will be given for acute conditions which endanger the life or the health of the individual, for correctable chronic conditions in adults which are preventing normal economic adjustment of the family, and for chronic conditions in childhood that interfere with the health or normal physical and mental development of the child. However, if the case is an emergency or the economic status of the family and their distance from home makes return impractical, further care, including obstetrical care, will be given.[22]

The AWHAs provided care to migrant workers in federal camps and those migrants outside the camps who were ineligible for county, state, or private assistance because of residency requirements. This included most migrant workers and their families. Physicians' groups again insisted that the FSA establish an income eligibility threshold, but the level chosen (net annual in-

come below $1,200) excluded few migrant families. The FSA tried to keep el-
igibility criteria as loose as possible. This left open the possibility that under
certain circumstances local residents might obtain medical care in the FSA
clinics. The FSA maintained that many "counties are not able to help these
families or do not have the desire to do so," and pressed for expanded coverage to indigent rural families and migrants whether they were camp residents
or not.[23] When the war made medical personnel scarcer, clinics did become
important sources of medical care for the local nonmigrant population, and
FSA nurses more frequently ministered to low-income farmers who were not
themselves migrants.

Once registered in a federal camp, individuals and families were issued
medical identification cards that ensured access to existing medical clinics located in federal farm labor camps throughout the state. Immunization status,
venereal disease serology records, and treatment records appeared on the back
of the cards. Full medical records rarely moved as quickly as the migrants
themselves, a source of frustration for both physicians and FSA personnel,
who were never quite sure whether the treatment provided to migrants had
been either completed or successful. Henry Daniels, who began as a camp
manager and later became the AWHA national supervisor by the end of World
War II, recollected that "one of the problems was that you would start a therapeutic session for somebody and they would move from there to another
camp or out of the camp system, or to another Agricultural Workers Health
Association. So that it was difficult to know what the outcome really was."[24]

The health of women and children occupied an important position in the
priorities of the FSA's experiment in rural health care, a characteristic the
agency shared with other New Deal social programs and earlier reform efforts.
The agency's adoption of specific standards of care for maternal and child
health—for example, insisting that a woman be attended by a physician during birth—represented an unusually explicit government effort to monitor
and modify medical practice for that period. Those individuals interviewed for
this project believed that the health of pregnant women and children was materially improved as a result of the FSA medical care program.[25] Anecdotal evidence found in the comments of a home management supervisor from San
Miguel, California, provides a clear contrast between the traditional birthing
experience of rural women and what was at least possible in the FSA programs:

> During the first year of the medical service plan, 31 confinement cases were
> cared for. Not one mother or baby died. Eighteen of these mothers were Span-
> ish women. Only three of these eighteen had ever had a physician attend them
> at birth; 6 of these mothers had lost babies at birth; 2 had each lost 4 ba-
> bies . . . but all 18 Spanish mothers have healthy FSA Medical Service babies
> now . . . It seems that our plan has had some effect already on the terribly high
> infant mortality rate in this county.[26]

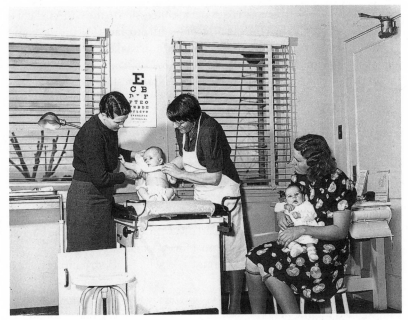

*FSA Administration, FSA Tulare Camp for migratory workers. Well-baby clinic.
Visalia, Calif., 1940. Leo Rosten.*

The FSA had to be pragmatic and flexible to gain the support of organized
medical groups at the state and local level. Participating physicians and medical societies had much of the responsibility for setting fee schedules and
defining which services were available in the program. Most participating
physicians considered the FSA program a necessity born out of extraordinary
but temporary circumstances, one that should be limited to the treatment of
acute medical conditions only. The FSA, in contrast, held to a more liberal
view that acute or chronic conditions that interfered with rehabilitation or active employment were justifiable expenses that should be covered. Eligibility
was a regular source of physicians' complaints as well. Local physicians and
county medical groups favored strict eligibility criteria and were displeased
when nonmigrants were allowed to use FSA medical services. Whenever controversy regarding eligibility arose, the agreements specified that they would
be resolved through discussions between camp managers (or sometimes regional or district FSA administrative officers), state and county relief agencies,
and individual physicians or representatives of organized medical groups.[27]

Because the FSA tried to accommodate local conditions and the concerns
of the local medical community, there was a fair degree of regional variation
in the program, not unlike that seen in the medical care cooperatives. In California, for example, migrants were more commonly referred to local physi-

cians' private offices. This was a reasonable concession to the CMA in light of its earnest support of the FSA's farm labor health programs. Across the border in Arizona, however, large distances between farm worker population centers and private doctors' offices resulted in fewer out-of-camp referrals. Similarly, the migrant health programs in the Pacific Northwest, Texas, Florida, and along the East Coast were characterized by fewer off-site referrals to local doctors.

Migrants typically followed well-defined harvest routes through contiguous states. Moreover, medical societies in the western states often shared educational conferences and medical journals. In several instances, the FSA tried to create interstate agreements with state medical societies to cover an entire region. Few state medical societies found the administrative and organizational advantages of regional agreements as compelling as did the FSA, however, and these efforts typically foundered.[28] Consequently, while the FSA administered the AWHAs as regional programs, separate agreements with each state medical society set forth the terms under which the program operated in any given state.

Nurses carried out much of the routine medical treatment in the migrant camps, but the participation of local physicians and the sanction of county and state medical societies were essential to the development of the program. Physicians recruited to the medical care program typically staffed two to three clinic sessions per week, usually on a rotating monthly basis. Recruitment was not a problem since the migrant health plans, like the FSA's medical cooperatives, provided needed income and relieved local physicians of the large seasonal burden of uncompensated care. As Karl Schaupp of the California Medical Association wrote in a 1944 article, the FSA program gave migrant families "a type of care that was entirely new to this group and was far better than anything they had ever known," and it allowed physicians "to practice a better type of medicine, and, for the first time, he received compensation for his service."[29]

In communities where doctors were either not willing to supervise nurses or provide medical care—or where there were simply too few physicians— the FSA did not hesitate to request support from the U.S. Public Health Service. As noted previously, the FSA's reliance on medical officers and public health nurses from the Public Health Service expanded during World War II, a fact that, as we shall see, came to have adverse political consequences for the medical care programs. U.S. Public Health Service medical officers were involved both in providing on-site medical care as well as supervising district or regional migrant health programs.[30] In addition to these duties, U.S. Public Health Service physicians implemented some of the more innovative policies adopted by the FSA migrant health programs, including policies that extended the influence of the program outside the camp's perimeter. In Arizona,

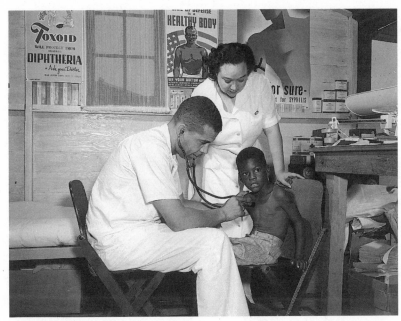

Farm Security Administration agricultural workers' camp. The clinic. Bridgeton, N.J., July 1942. John Collier.

for example, a district Public Health Service medical officer decided not only to routinely immunize all federal camp inhabitants, but sent the camp nurse out to employer-owned camps and squatter villages on an immunization campaign. This policy, like the policy of including dues in annual rehabilitation loans, emerged from the field. The FSA later adopted it as a national standard and tried to implement it throughout the migrant health plans.[31]

A wide mix of medical ailments were handled by the AWHA clinic staff. Upper respiratory and gastrointestinal complaints topped the list of diagnoses (Table 3.1). Communicable diseases were frequent, including tuberculosis, typhoid, and diphtheria.[32] Occupational diseases also drew more attention from both the federal government and trade unions during the 1930s and 1940s. In the migrant health programs, dermatitis and pesticide poisoning from lead arsenate, as well as serious trauma and a variety of musculoskeletal complaints, were common.[33]

Isolation from families, crowded living conditions, poverty, and readily available alcohol and sex in "juke joints" made venereal disease endemic wherever migrants congregated. Treatment for syphilis and gonorrhea at the time was possible, although penicillin had yet to supplant the painful arsenicals then in common use. The FSA collaborated with local and state health departments (and during the war with the U.S. Army) to identify and treat vene-

TABLE 3.1 DISTRIBUTION OF ILLNESSES IN THE AGRICULTURAL WORKERS HEALTH AND MEDICAL ASSOCIATION, 1939–1940

Diagnosis group	All ages	Age groups									
		0–4	5–9	10–14	15–19	20–24	25–34	35–44	45–54	55–64	>64
Disorders of											
Digestive system	26.3	19.1	22.9	23.7	18.7	21.9	32.8	34.8	34.5	30.6	36.7
Respiratory tract	21.1	29.9	35.3	27.4	17.8	12.8	18.6	13.9	16.8	11.9	13.3
Pregnancy and childbirth	10.3	—	—	0.8	30.5	32.1	12.6	8.0	0.4	—	—
Genitourinary system	6.1	4.0	3.9	3.4	8.4	8.1	6.9	6.5	6.5	9.7	6.7
Accidents	9.2	11.9	11.4	15.0	6.8	8.8	5.9	10.0	6.5	7.5	—
All other diagnoses	27.0	35.1	26.5	29.7	17.8	17.2	23.2	26.8	35.3	40.3	43.3
All disorders	100.0	100.0	100.0	100.0	100.0	100.0	100.0	100.0	100.0	100.0	100.0

Source: Report of Agricultural Workers Health and Medical Association, June 1941, Record Group 224, Box 137, National Archives, Table 6, p. 24.
Note: Based on an approximately 10% sample of AWHMA referral data for the fiscal year 1939–40.

real diseases, but these efforts were made difficult by the migrants' mobility and by racial and ethnic stereotyping.[34]

In California and Arizona, office visits cost the FSA, on average, $1.50, while clinic visits averaged only $1.35 per patient.[35] Compensation rates for physicians averaged $7 for the first hour of clinic and $1.25 for every quarter hour thereafter. Emergency home visits were reimbursed at $2 a visit, with a mileage surcharge. If a doctor saw more than one family on any emergency, he or she received an additional $1 per family.[36] Physicians staffed evening clinics to increase their availability to migrant workers, the majority of whom toiled in the fields until dusk. Nurses and physicians alike worked long hours, and it was not unheard of for upward of thirty patients per hour to be seen. This prompted many participating doctors to complain that they were forced to provide poor-quality care, and the FSA Health Services Branch worried that overworked physicians would be less supportive of the program.[37] When the illness was serious, or if on-site physician care was unavailable, camp residents were referred to local physicians' offices for treatment. This required authorization by the camp nurse or manager, an example of the administrative control exercised by the government in the migrant health program.

Dental care was not a major focus of the FSA medical care programs, but neither was it ignored. As far back as World War I, when the Selective Service experience revealed the poor baseline health status of rural recruits, it was well known that rural families had abysmal teeth and gums.[38] While acute dental emergencies might force farm workers to seek care, preventive visits to the dentist were rare. In the federal camps, migrants were able to obtain acute or emergency dental care through regularly held dental clinics or (less commonly) through referral to local practitioners. Eligibility criteria were again loosely defined, and children were provided with both preventive and emergency care. [39] Although as many as 700 people per month might be seen in the dental clinics of many federal camps during peak harvest time, they sometimes needed a bit of gentle persuasion to visit the dentist. "When business was slow," a Yuba City, California, camp manager reported, "we would announce the dentist was leaving at the end of the week, and in 10 minutes, the waiting room was full."[40]

The annual cost of funding the migrant health program was $2 million divided between the agricultural workers health associations. This averaged out to be between $1.50 and $2 per migrant per month.[41] Three-quarters of total costs went to pay for the care provided by doctors, hospitals, nurses, and dentists. A surprisingly large portion of the budget, 14 percent, went to nutritional supplements for the children in the camp schools and day-care programs. Administrative expenses consumed 13 percent of expenditures, a figure that held reasonably constant in the various plans.[42]

The FSA or the AWHAs themselves contracted with area hospitals to provide hospital services at reduced rates. The agricultural workers health associations set limits on length of hospital stay (typically fifteen days), and created mechanisms for medical societies or FSA staff physicians to review and audit hospital stays. Preauthorization by physicians was required for hospitalization, except in an obvious emergency when nurses or camp managers could make the decision alone. If hospitalization was complicated, exceeded fifteen days, or if unique treatment, testing, or appliances were required, second opinions from medical consultants were mandatory.[43] Another administrative cost control was that only those medicines listed in the *U.S. Pharmacopoeia* or *National Formulary* were paid for in the migrant health program. These administrative policies represented restrictions on medical practice that were unusual for this era. For example, the formulary policy had a larger impact in rural communities because of their more frequent use of folk remedies and because many rural physicians supplemented their income with their own mortars and pestles.[44]

The FSA took the highly unusual step of purchasing two hospitals, becoming the first federal agency outside of the U.S. Public Health Service to own and operate hospitals for nonmilitary personnel. In 1941, with the acquiescence of the Arizona Medical Association, the Agricultural Workers Health and Medical Association purchased the Burton Cairns Convalescent Center in the Eleven-Mile Corner area, transformed it into the Burton Cairns General Hospital, and ran it until July 1944.[45] Local physicians had full admitting privileges to the hospital and access to all of its acute-care facilities, including a surgical suite, an emergency room, and in-patient units. Associated with the hospital was what would now be called an intermediate care facility where patients not yet prepared to return to their home could be sent to recuperate. Burton Cairns General Hospital was used extensively during World War II for agricultural workers brought to the United States under treaty agreement with Mexico, and pregnant wives of infantrymen in the first four pay grades gave birth there as well under provisions of the Emergency Maternal and Infant Care (EMIC) legislation.[46] Beginning in 1942, the FSA operated a second hospital for the farm workers in south-central Florida. The Migratory Labor Hospital in Belle Glade was owned by the Migrant Labor Health Association, but its funding came entirely from the FSA. The two hospitals enabled the FSA to improve access to medical facilities for its migrants in communities sorely lacking in such facilities. In the case of Belle Glade, this problem of access was compounded by racial attitudes that strictly limited the access of nonwhite residents to existing facilities.[47]

Harold Mayers, hospital administrator in Belle Glade, remembered that the FSA was sometimes caught between local doctors who wanted to limit eligibility, and community residents who felt that they ought to have access to the same facilities as the migrants. "There was a constant fight between migrants and the people," recalled Mayers. "They would say, 'This is a beautiful hospital, we live here, we pay all these taxes to the federal government, but we can't go out there and get any kind of care.'"[48] The FSA tried to accommodate local citizens whenever possible, and it was more than willing to let nonmigrant families use its medical facilities. However, the FSA's openness on this issue could and did put the agency in conflict with the local medical establishment, which was vigilant against what it perceived to be unfair competition from the federal government.

PRETTY LONG HOURS AND PRECIOUS COOKERS: THE ROLE OF NURSING AND HOME MANAGEMENT

The leadership of the FSA Health Services Branch believed in the power of prevention and education to break long-standing patterns of disease in rural areas. This commitment took many forms, from teaching women how to can fresh vegetables to promoting elected camp councils to adopting rules governing sanitation and hygiene inside the federal camps. Preventive measures such as childhood immunization, nutrition programs, improved sanitation, prophylactic dental services, and case finding for communicable diseases (i.e., the process of locating and evaluating individuals who have been exposed to a diagnosed case), were viewed as essential adjuncts to the more traditional emphasis on the diagnosis and treatment of disease.[49]

Nurses and home management supervisors were the linchpins of this effort. Although the FSA utilized nurses and home management supervisors in some capacity in all of its medical care programs, their roles were fundamental to the migrant health program. These women (and they were all women) shared the vision of public health advocates of the era and viewed their work as the start of a national program for all of society's disadvantaged families. "It is a long and far-reaching program which has been undertaken," wrote an FSA nurse to her colleagues in the *Pacific Coast Journal of Nursing*. "At times the discouragements are great and the apparent rewards few. However, it is a step in the right direction and those connected with the program are looking forward to the day when adequate medical assistance for the agricultural migratory worker will have come to mean adequate medical assistance for all low-income groups."[50]

The FSA was not the first to use nurses in expanded clinical roles. Nurses were given clinical responsibilities in the military and in various rural demonstration projects funded by philanthropies during this period, and they were an important professional element in the urban settlement house movement of the Progressive Era. Some public health departments used nurses in similar roles.[51] However, the degree of clinical and administrative responsibility given to women in the FSA was unusual for the time. Women were placed in second- and third-tier administrative positions in the FSA's home management and nursing divisions. Whether the New Deal's approach to women was intended to break established social norms, or whether women were assisted as an incidental consequence of its myriad domestic social welfare programs is debatable. In the FSA migrant health programs, the advancement was purposeful. Nurses' central clinical, administrative, educational, and public health responsibilities in a publicly funded civilian medical care delivery program that operated on a national level distinguishes the FSA programs from other initiatives. "We gave those nurses as much authority as we could possibly give them," remembered Lorin Kerr, "and still get away with it." [52]

In the migrant health programs, Farm Security Administration nurses were given unusual latitude in their clinical roles. They conducted health screenings for new camp residents to identify conditions that were "likely to endanger the health of other camp occupants, or chronic conditions likely to render the applicants ineligible for camp residence."[53] With the verbal approval of the camp doctor, they could write prescriptions and dispense drugs from the clinic formulary. Nurses filled other roles as well. They staffed well-baby clinics, coordinated immunization programs, and used the camp's recreation facilities and schools for health education and nutrition classes for men and women. Expectant mothers were given regular prepartum examinations, and nurses tracked them after delivery to make certain they and their new babies received proper postpartum care. Nurses visited squatter camps and, whenever possible, private growers' camps to encourage pregnant women to come to educational programs or the clinic for prenatal or postpartum care.[54]

Nurses decided whether a sick migrant required referral to a local physician's office or to the hospital. They handled a variety of routine health complaints, provided minor medical and surgical treatment, and monitored physicians' treatment plans. If a physician was unavailable in an emergency, nurses were on hand to carry out triage and to provide emergency care.

Nurses were communitywide emissaries for the FSA as well. They collaborated with other state and federal agencies on various health-related issues and they nurtured the FSA's relationship with local parent-teacher organizations and Red Cross chapters, county professional societies, and public health and relief agencies. "They went out and worked with the doctors, they worked with the hospitals, they worked with the people," recalled Kerr. "They were fantastic and they were on the go all the time!"[55] Henry Daniels added that there were political as well as practical advantages in creating a program in which nurses occupied such a pivotal position. "I think it was easier to go into the communities with nursing service and utilize the physician service that was there," Daniels commented, "rather than trying to superimpose a delivery system that was based upon full-time salaried MDs. It was more acceptable."[56]

TABLE 3.2 TYPES OF SERVICES PROVIDED BY AGRICULTURAL WORKERS
HEALTH ASSOCIATION CLINICS, JUNE 1944

Type of service	Total
Examinations	3,903
Immunizations	1,911
Dental care	1,569
Cases of illness, including venereal disease	6,264
Total cases	13,647
Visits	
For illness	
To physicians	12,353
To nurses—home	8,891
To nurses—clinic	26,149
Total	47,393
For examinations	4,905
For immunizations	2,605
Dental visits	2,923
Total visits	57,826
Visits	
To physician per case of illness	2.0
To clinic nurse per case of illness	4.2
To clinic nurse for all medical cases	2.9
Cost	
Per visit, all visits	$1.66
Per case, all cases	$7.06
Total clinic costs	$96,387.23

Source: U.S. Department of Agriculture, Farm Security Administration, Health Services Provided through
Agricultural Workers Health Associations, 1944, Table 3.

Nurses filed monthly reports with the FSA's district and regional offices.
Their diverse responsibilities are evident from the briefest scrutiny of these
reports. These monthly reports also document the evolution and expansion of
nurses' roles and they underscore the FSA's emphasis on education and prevention. These reports enabled the FSA to track the type and volume of services provided in the camp clinics, including the number of clinic and home
visits and hospitalizations, as well as the types of illnesses seen by camp nurses
and physicians (Table 3.2). Pauline Koplin, who worked as an FSA camp nurse
in Yakima, Washington, remembered that "every day was different. One day
might be set up for well-baby clinics in the morning and general [clinic] in the

afternoon . . . Sometimes it was immunization day. We'd start our day about seven in the morning and finish about twelve or one at night. It was pretty long hours."[57]

Camp nurses were supervised by participating local doctors or by U.S. Public Health Service medical officers assigned to the camp clinics. District and regional nursing supervisors in the FSA's Nursing Branch maintained close contact with the agency's field nurses, as did the district and regional medical officers who were usually assigned to the FSA from the Public Health Service. One of the duties of the FSA's physician staff was to intercede on behalf of camp nurses whenever problems arose, especially if they involved local doctors or county medical groups. This parallel chain of nursing and medical supervision extended to the Health Services Branch office in Washington.

Nurses used standing orders issued by the FSA's district or regional medical offices and approved by local physicians. The orders served both a protective and an instructive function for nurses, enabling them to treat most routine medical illnesses and providing guidance about how to proceed in those instances when physician backup was unavailable or delayed. "Nurses functioned pretty autonomously," Henry Daniels later recalled. "They were able to do a lot of what nurse practitioners do after a lot of training, but these nurses did it through experience."[58] Although local physicians sometimes expressed consternation about the nurses' role in clinical decision-making, the agency's physician staff generally found nurses to be reliable diagnosticians and skillful in their management of patients.[59]

The FSA distributed handbooks and manuals to its field staff that gave guidance on a range of issues they were likely to confront. A *Community Nursing Handbook,* for example, included meticulously detailed appendices crammed with information on a wide variety of topics. It contained instructions on stocking the camp infirmary and clinic, constructing examination tables from raw materials, preparing self-contained "Home Delivery Service" packages for emergency deliveries, as well as suggestions for reducing the threat posed by disease vectors.[60] It is hard to deny that the agency's educational efforts were laced with paternalism. In a 1941 article carried in the *Pacific Coast Journal of Nursing,* one FSA nurse warned her colleagues to expect "ignorance and superstition in occasional cases which will require tactful and careful handling." "Only with patient and careful watching," she continued, "does the mother's faith become established in nose drops and camphorated oil rub and does [she] finally accept the fact the 'Black Draught Purgative' must not be given to any family member if she has a pain in the abdomen."[61]

Some nurses had the same difficulty adjusting to their diverse roles as did local doctors. The result was sometimes confusion and unease, and sometimes acrimony and controversy. This tension was frequently couched in terms of safe and ethical medical practice. However, deviation from normative social and professional roles played no small part. There were leaders in the nursing community, for example, who advised the FSA that nurses might inadvertently overstep legal, ethical, and clinical boundaries. Olive Whitlock, director of the Division of Public Health Nursing for the Oregon State Board of Health, set out her concerns in a letter to the FSA regional office in 1939. "It seems the nurse is to see all patients requiring medical care, and refer them to the contract physician if she is unable to give the necessary care herself . . . My chief concern was over the amount of responsibility placed upon the nurses for determining the need for medical care and dispensing simple 'remedies.'"[62]

The FSA's Home Management Branch was, along with nursing, at the center of the agency's migrant health programs. Home management supervisors had their own handbooks with details on planning and preparing meals, sanitation, porch and window screening, and canning and pickling home-grown fruits and vegetables. As described in the *Handbook for Camp Home Management Supervisors,* their duties were to "carry out a basic health education program that will be uniform throughout the region, develop community consciousness and solidarity and contribute to the health, welfare and happiness of the migrant workers families."[63] Women and children were the prime target of the home management staff. Home management supervisors coordinated school and day-care lunch programs. Mothers were admonished to monitor the weight of their children because an "optimal diet for each and every child" would improve school performance, health and nutritional status, and family well-being. No detail was too small to pass along. One handbook suggested that for lunch "egg nogs, milk, chocolate, chocolate milk, all kinds of fruit juices, or fresh fruit may be combined with a sandwich or cookies" and using garnishes would "keep favorite dishes favorite."[64]

A basic philosophy that underlay the FSA medical program's educational efforts was that low-income rural families should assume greater responsibility for their own lives. To this end, the FSA's health education programs aimed to provide specific knowledge, to inculcate self-sufficiency, and to promote personal responsibility in matters of health and hygiene. The agency anticipated that as families became more informed and confident, they would be less of a burden for the busy medical staff. To help meet this goal, the FSA distributed pamphlets with titles such as "Care of the Sick," "Keeping the Baby Well," "Facts about Cancer," "Preparing for Marriage," "Step by Step in Sex Education," and "From Seeds to Saplings." Materials on building privies, installing screens, and protecting home water supplies were widely distributed as well.[65]

Nurses and home management supervisors formed "mothers health clubs" whose purpose was "to teach all mothers the value of health education in the home, how to improve environmental conditions that [a]ffect mental and physical health, and the importance of prenatal, infant and child care." Clubs adopted mottoes (e.g., "Give to the world the best you have and the best will come back to you," "Health is a way of living, rather than a subject to be taught") and provided information on medical, nutritional, and sanitation issues to the newsletters that camp governing councils sometimes published. These councils formed their own health committees as well and gave them educational and policing duties in the farm labor camps, such as monitoring the attendance of pregnant women at prenatal classes.[66]

Sometimes the FSA's educational efforts in family planning and sex education in the farm labor camps caused both controversy and mirth. T. G. Moore recalled that when he brought in one of Margaret Sanger's protégés to a resettlement project in Robstown, Texas, he scandalized a few families and came in for a bit of good-natured ribbing by some of the camp women:

> They sent two nurses out with a training program with their rubber bodies and they brought with them several big boxes of foam powder and sponges. I knew I was treading on water, but I wasn't quite prepared for what happened. Anyway, you used water to shrink the sponge filled with foam and work it up and then you insert it. I remember one woman who had eight kids . . . I'll never forget her, 'cause she said, "My God, I'd be pregnant before we'd get the windmill turned on." And that's a true story, because you had to go outside for water, there was no water in the house! [67]

No program demonstrated a more seamless liaison between the U.S. Public Health Service and the FSA than the nutritional efforts coordinated by the FSA Home Management staff. The Public Health Service's involvement with the rural southern diet predated the FSA by some twenty years through the pioneering, if much-neglected, work of Joseph Goldberger. In 1916, Goldberger undertook a study of the biologic, social, and economic factors that caused pellagra, then a major cause of morbidity and mortality in the South. Aided by the same Edgar Sydenstricker who figured prominently in the CCMC and the Technical Committee on Medical Care, Goldberger identified a missing dietary element as the cause of pellagra. Later researchers proved that the missing element was niacin but, in a triumph of empirical science, Goldberger and his colleagues demonstrated that simply broadening the diet prevented the disease. In one of the least acknowledged but most stunning successes of applied science and public health, the combined efforts of numerous public and private agencies eradicated pellagra by the early 1940s.[68]

The FSA institutionalized Goldberger's work in its nutritional programs and is credited for its substantial role in the successful fight against the niacin-deficient southern diet. As part of this effort, home management supervisors instructed women in the migrant camps on the nutritional benefits of home gardens and how to can the fresh vegetables and fruit they grew. Home management supervisors handed out pressure cookers (called "precious cookers" by many of the camp women) to preserve the nutrients in vegetables often lost in cooking, and emphasized that dietary diversity prevented illness and nurtured good health.[69] Just as FSA supervisors taught the men how to diversify their crops and plant home gardens, home management supervisors and nurses taught women how to cook and can the fresh produce grown by the family.

The FSA collaborated on many fronts with various federal, state, and local public health agencies, as the pellagra example illustrates. The FSA's sanitation engineering staff, for example, participated in part of the larger New Deal effort to improve sanitation in rural communities. This was a monumental if not always well-coordinated effort that combined the collective resources of the FSA, the Works Progress Administration, the Civilian Works Administration, the U.S. Forest Service, the Civilian Conservation Corps, and the U.S. Public Health Service, as well as many local and state health departments. Between 1933 and 1942, wells were sunk to protect family water supplies; privies were built; and windows and porches were screened to keep out mosquitoes, flies, and other disease-carrying vermin throughout rural America.[70]

PHYSICIANS' REACTIONS TO THE FSA MIGRANT HEALTH PROGRAM

The creation of the first agricultural workers health association initially touched off a series of testy exchanges between the AMA and the California Medical Association, some of which were carried in the minutes of the AMA's annual meeting. However, the AMA eventually adopted more of a wait-and-see attitude. This can be inferred from the publication in June 1938 of an article in the *Journal of the American Medical Association* by Dr. Walter Dickie of the California Medical Association in which he outlined the goals and anticipated impact of the Agricultural Workers Health and Medical Association. The favorable remarks made by a California general practitioner at the CMA's 1938 annual meeting were also carried in a 1939 edition of the *Journal of the American Medical Association*. Speaking about the FSA camp at Riverside, Dr. Mary Baldwin commented that "it was immaculately clean . . . The nurse was doing a tremendous piece of health education work. She had literature in a little waiting room on all the subjects we want people to know about: syphilis, maternal health, all that sort of thing."[71] Comments on the migrant health program in the *Journal of the American Medical Association* and at the AMA's 1940 and 1941 annual meetings generally offered no substantive criticism.

At the local level, physicians' support for the migrant health programs was driven largely by the same factors that nurtured the medical care cooperatives.

Demand for uncompensated care was more severe in communities burdened by a large influx of migrant workers, and local physicians were usually quite appreciative when the government assumed the financial burden that had heretofore been theirs. "Local physician groups felt some sort of satisfaction and gratitude," Henry Daniels recalled, "because somebody was coming in and trying to do something about a problem that they had not been able to handle."[72] The FSA's flexibility in adapting to local concerns and the agency's policy of close cooperation with county and state medical societies certainly remained influential considerations with local doctors and organized medical groups as well. Despite the racism and xenophobia that the professed fear of contagion often masked, the profoundly unsanitary living conditions of many migrants evoked justifiable concern about public health on the part of physicians and the community at large, and there was genuine relief when these were addressed by the FSA. As Karl Schaupp commented in a 1944 *California and Western Medicine* article, the migrant health program "protected the health of the citizens of the localities served because to a large degree it controlled the spread of epidemics which before had followed the migratory worker."[73]

The FSA farm labor program had an impact on disease patterns in the communities where it operated. Acting in conjunction with local, state, and federal authorities, the agricultural workers health associations moderated and probably prevented outbreaks of communicable diseases. In 1939, for example, California experienced a series of devastating floods. Fearing a typhoid epidemic outbreak, state health authorities, the FSA, and physicians working for the Agricultural Workers Health and Medical Association mounted an aggressive campaign to inoculate some 75,000 migrants against typhoid and smallpox. The anticipated epidemic never materialized, and the death rate from typhoid fell to record lows.[74]

Participating physicians and cooperating medical societies credited the AWHAs for shouldering a welfare burden that no one wanted to assume, for the economic benefits that accrued to physicians, and for the FSA's efforts to work collaboratively with the medical profession. Schaupp wrote that the Agricultural Workers Health and Medical Association protected communities while giving physicians the opportunity to provide care to a group that historically received inadequate medical attention and for whose care doctors had rarely been paid. For example, 50 percent of the AWHA's funds, on average, went directly to physicians. A physician member of the board of directors of the Texas Farm Laborers Health Association reported to fellow board members that doctors were "all fairly well satisfied. One reason—of course—is because they are getting paid and because of the fact that they feel they have a little voice in it."[75]

Interestingly, some doctors valued the cooperative venture because it gave physicians and medical societies—many for the first time—experience with third-party payers at a time when medical practice was still dominated by fee-for-service. Toward the end of the 1930s and during the war, there was intense discussion about health reforms in medical and political circles. As third-party health insurance, in the form of hospital insurance plans like Blue Cross or medical service bureaus, grew in popularity, some physicians found new reasons to support the FSA's medical care programs. "There is one other important feature about this work which should be generally known," noted Schaupp. "In this piece of work we are getting that experience which will be of great importance in any future planning of health insurance."[76]

Although the FSA migrant health program was received favorably in most communities, the medical profession was not then—nor perhaps ever—of one mind about any reform proposals. The migrant health programs were no exception. Both individual doctors and organized medical groups vacillated in their attitudes toward the agricultural workers health associations. An FSA lawyer charged with developing the contract language to be used in the memorandum of understanding in the Northwest complained to his supervisor that "we, who have been working on the medical care program in the region, have experienced this same attitude repeatedly; that is, an apparent comprehensive understanding and a cooperative attitude in one instance and, at a later date with the same group, a confused understanding and a change of attitude."[77]

In communities where the migrant health clinics operated, physicians also raised questions about the program's loose eligibility criteria, low reimbursement rates, and real and perceived encroachment into matters best left in the hands of doctors—none of which were new to the Farm Security Administration. Local physicians occasionally rebelled at the use of standing orders and the clinical latitude enjoyed by FSA nurses. Conflicts at the local level between nurses and physicians may have been due to simple personality clashes or legitimate and genuine concern with each other's clinical competence, or they may have been emblematic of the tensions caused by a program that stretched the usual social and professional roles. What is certain, however, is that they did occur with some frequency.

In 1940, a nurse working in Idaho was transferred to another farm labor camp because local doctors found her "domineering" and accused her of taking "too positive a stand as to whether treatment should be given certain of the patients and in some instances as to what treatment should be administered."[78] In 1939, a vexed surgeon in Washington State chastised Chief Medical Officer Ralph Williams for "permitting nurses to give medications . . . and for assuming the responsibility for deciding when the patient needs medical care."[79] Williams replied quickly, informing the physician that "it is not the policy of this Administration to permit nurses to practice medicine. We are thoroughly aware of the dangers of permitting any nurse to assume too much responsibility in connection with the treatment of disease. We have been very punctilious in this respect relative to the nurses employed in the Migrant Labor Camps that are conducted by this Administration."[80] Protestations on the part of the FSA medical staff that camp nurses always worked under close supervision and the use of standing orders signed by the local physician did not always assuage doctors' concerns. Moreover, some inside the FSA suspected that these objections reflected physicians' fear of competition rather than genuine concern for clinical competence.[81]

The minutes of the board of directors meetings for both the Texas Farm Laborers Health Association and the Agricultural Workers Health Association in the Pacific Northwest reflected repeated criticism by physicians of financial issues, such as fee schedules and delinquent reimbursement.[82] As in the medical care cooperative plans, some doctors charged above the established fee schedule and others wrote costly or unnecessary prescriptions. Physicians occasionally classified minor surgical procedures as major operations or recommended unnecessary surgery. At a farm labor camp in the state of Washington, for example, an entire family of eight was scheduled for tonsillectomies by an especially eager surgeon. (They escaped the scalpel by departing in the middle of the night.) The FSA saw these as efforts on the part of participating doctors to pad their incomes, a concern raised at times by physicians as well.

Even in supportive state and county medical societies, such as the California Medical Association, there was dissent over the FSA plans. Some physicians stated unequivocal opposition to federal intrusion in delivery of medical care, regardless of the circumstances that had led their medical colleagues to support the migrant health program. It was "desirable for the profession to continue to carry the lead in the care of the indigent," a doctor argued in a 1938 article, "rather than sacrifice the time-honored privilege of being the custodians of the health of the people."[83] Others challenged the AWHA's inroads into professional independence, as exemplified in the use of standing orders for nurses and the administrative restrictions placed on specialist referrals, hospitalization, and drugs.

In 1937, a group of physicians opposed to the migrant health program presented a resolution at the CMA annual meeting that would have had the society go on record as stating its "apprehension over the recent increasing tendency of a large group of our citizens to shift the responsibility for their present and future welfare upon a willing and paternalistic government."[84] The resolution was defeated, but only after Karl Schaupp (who noted that his committee was "highly sympathetic") suggested that it was politically imprudent given that the state's governor was pushing for passage of a state-based compulsory health insurance program.

This controversy illustrates the dilemma facing doctors and the medical community as they contemplated the FSA programs, a predicament that was captured in a 1937 editorial in *California and Western Medicine:* "The federal government is cooperating with the State in trying to work out a solution," the article noted. "It is hoped that the component county societies . . . will work hand-in-hand with the representatives . . . for the protection of local interests and to make unnecessary the inauguration of measures partaking to [sic] the nature of state medicine."[85] Clearly the medical profession's approval was in part driven by the consideration of less palatable proposals in the political arena. For many physicians, cooperation was less an expression of heartfelt support than it was a realization that the FSA program was like a dose of the medicine they often prescribed for their patients: it might not taste good going down, but it was better than the alternative.

CONGRESS INVESTIGATES THE AGRICULTURAL WORKERS HEALTH AND MEDICAL ASSOCIATION

Beginning in the 1940s, a series of congressional investigations into the FSA introduced a period of increasing public scrutiny of the agency's activities.[86] The medical care programs were not initially the focus of these inquiries; nevertheless, they came in for some harsh criticism as the FSA battled for its survival. Congress initiated two detailed audits of the Agricultural Workers Health and Medical Association during this period of increasingly partisan political debate over the New Deal's social agenda. The audits illustrate features of the FSA migrant health program that presaged innovations in health care delivery often considered more contemporary, such as restrictions on specialist referrals, hospital stays, and the use of nurses and general practitioners to control access to these and other services. In addition, the audits offer evidence of the historical roots of some of the more vexing and intriguing phenomena found in medical care delivery in the United States. It is for the purpose of exploring these issues that the audits are considered here while the political environment is discussed in detail in subsequent chapters.

The first investigation was begun in 1940. It was conducted by William Goldner, an independent auditor who was employed by the Bureau of Agricultural Economics. His report, "The Costs of Medical Care in the Agricultural Workers Health and Medical Association" (a title that was clearly evocative of the Committee on the Cost of Medical Care) provides a wealth of detail on the operation, administration, costs, and services of the largest AWHA in California and Arizona. Goldner analyzed operating expenses for the Agricultural Workers Health and Medical Association for the fiscal year 1939–40 using a random 10 percent sample of all membership numbers to estimate prevalence of illness and the costs associated with these illnesses.

Goldner noted a number of administrative problems, but on the whole he was rather gentle in his critique. He decided that the flexible eligibility policy that irked many physicians was necessary and "prevented families barely above the relief level from being pushed back onto the relief rolls by the burden of a large medical bill." Goldner strongly supported the humanitarian objectives of the association and credited it for "setting the standard of adequate medical care for the migrant population in California and Arizona."[87] Goldner was not alone in this assessment; it was widely shared by the California Medical Association and pro–New Deal politicians and newspapers.

Goldner was an accountant by training, and his interest was piqued by several intriguing differences between the two states and even within states in terms of where services were provided, who received care, and wide variations in diagnoses made and the costs associated with specific diagnoses. On average, for example, 66 percent of all visits occurred in doctors' offices. In California, however, this figure was over 75 percent, while in Arizona it was closer to 47 percent. Goldner pointed out that a small percentage of eligible families received a disproportionate amount of medical care and generated higher costs per illness as well. Ten percent of all eligible migrant families consumed over 40 percent of total costs, Goldner noted, while 54 percent of eligible families utilized just 10 percent of total expenses.[88]

More remarkable to Goldner was the fact that diagnoses, type of services provided, and the fees charged per diagnosis showed curious variations between the two neighboring states and even within each state. In California, major surgical procedures accounted for 17 percent of total direct medical care expenditures. In Arizona, this figure was 23 percent. A similar inconsistency was seen with obstetrical fees. Office visits, however, accounted for almost identical percentages of total expenditures in both states: 14.5 percent in California and 12.4 percent in Arizona. Goldner felt that this between-state variation was not due to actual differences in the diseases and severity of illness in a group that traveled easily across state borders and that he considered homogeneous in terms of socioeconomic and baseline physical status. Even within similar diagnostic categories, Goldner documented significant cost variations. In the top fifteen diagnostic categories, for example, average expenditure per patient per month was almost $7 more in Arizona than in California.[89]

Goldner reanalyzed his information, restricting his analysis to diagnoses "which by their very nature must include exactly comparable amounts of service."[90] In this category he included tonsillectomies/adenoidectomies and stillbirths/miscarriages/abortions. The variability remained. He tried it again, this time accounting for differences in volume between the two states, and still the difference remained (Table 3.3). Goldner puzzled mightily over his findings, finally concluding that perhaps differences in complexity and frequency of referral of patients to physicians' offices, or differential use of tech-

TABLE 3.3 CALIFORNIA AND ARIZONA COSTS BY DIAGNOSTIC CATEGORY
IN THE AGRICULTURAL WORKERS HEALTH AND MEDICAL ASSOCIATION,
1939–1940

Diagnosis	Average expenditure per patient-month		Average expenditure per hospital day	
	Arizona	California	Arizona	California
Live births	28.69	46.33	5.22	6.11
Appendectomies	44.39	60.86	5.09	6.12
Pneumonia	39.85	36.36	3.98	4.79
Tonsillectomies and adenoidectomies	11.19	13.13	6.98	10.80
Diarrhea, enteritis, and colitis	32.68	33.95	3.55	4.71
Nephritis, other kidney disorders	38.99	49.75	3.86	5.80
Complications of childbirth and pregnancy	30.61	30.82	4.48	5.24
All other diagnoses	35.63	40.38	4.02	5.05
All diagnoses	32.31	40.21	4.26	5.62

Source: Report of Agricultural Workers Health and Medical Association, June 1941, Record Group 224, Box 137, National Archives, Table 21, p. 49.

nologies and consultation, might account for the phenomena. When he postulated that "a difference in the type of medicine practiced on the members" might explain these curious results, Goldner provided an early articulation of "small areas variation," a topic of intense policy interest in the late twentieth century.[91]

Ralph Gregg, a U.S. Public Health Service physician, conducted a second review of the Agricultural Workers Health and Medical Association in 1941. Gregg visited every clinic in Arizona and California, and met with the association's board of directors, FSA district and regional officers, California Medical Association representatives, and state health department officials. Gregg's stated purposes were threefold. First, to examine administrative procedures in the association. Second, to study the medical, dental, and nursing services, paying particular attention to financing methods and quality of service. And finally, to determine how well the statistical unit upon which the association based its internal auditing of the program's operation functioned.[92]

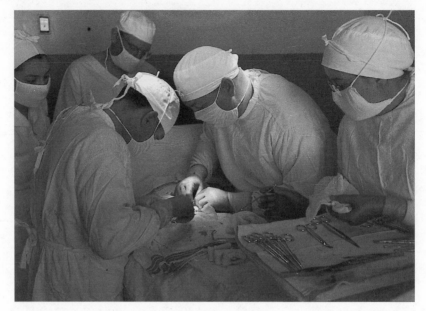

Operation in the Cairns general hospital at the Farm Security Administration migratory farm workers' community. Eleven Mile Corner, Ariz., February 1942. Russell Lee.

There were also political motives that led to this second investigation. In the first place, there was growing concern in Washington with international developments, and politicians in both parties were intent on tightening up on the government's purse strings. In a related development, the economic boom that occurred as the nation became the "arsenal of democracy" made it more difficult to justify continuing many New Deal social welfare programs. Finally, congressional opponents of the agency were more than ready to dig for dirt to use against the FSA, regardless of how parsimonious or profligate the agency was with its resources.

In his report, Gregg took aim at what he considered to be the association's unnecessarily lax organization and administration. He expressed skepticism of Goldner's earlier analysis because it relied on statistical data provided by the association's field offices, information that in Gregg's view was "cumbersome, confusing and unreliable."[93] Gregg believed that actual costs were considerably higher than those quoted by the association in its annual and monthly reports. In contrast to the approach used by Goldner, Gregg calculated clinic costs by including all salaries, travel expenses for nurses, clerical costs, material and supplies, rent and equipment costs, and physician and druggist fees in an aggregate sum. Gregg did not separate out the referral costs from the clinic costs per se, since many clinics also operated as referral centers and he maintained that teasing out proportional costs was problematic. Referring to statistics on Burton Cairns, for example, he wrote that the Agricultural Workers Health and Medical Association stated that the cost per bed-day was $1.90, while his calculations showed that the real costs were closer to $3 per day.

In spite of his methodological criticisms of the Goldner audit, Gregg identified many of the same issues in his review as Goldner had. Participating physicians were positive about the migrant health plan, but a number of them told Gregg that the sheer patient volume sometimes forced them "to see patients too rapidly to be able to practice good medicine." During the busy summer season in Chandler, Arizona, physicians amassed some seventy-seven hours of clinic work each week, often seeing up to thirty-two patients per hour. Gregg noted the long clinic hours worked by nurses and doctors, and he recommended that the association maintain its original goal of twelve patients per hour and increase a doctor's reimbursement according to the actual time worked. "Patients cannot be given adequate attention," admonished Gregg, "if the physician is required to see 20 to 30 patients per hour."[94]

Gregg demonstrated to his satisfaction that visits to private physicians' offices were generally more costly than clinic-based care. He recommended that salaried, clinic-based physicians be substituted for off-site physician referrals. "The clinics generally are as well-equipped as the physicians' offices," he commented, "and can serve as well in caring for most ambulatory conditions." Gregg believed that some physicians provided "good clinic work" to "induce patients to select them when they are referred out of clinic."[95] A related concern of Gregg's was that part-time physicians were too eager to refer patients to their private offices for follow-up care, especially in maternity cases.

The medical director of the Agricultural Workers Health and Medical Association, A. E. Larsen, took issue with the tone and content of Gregg's report. Larsen was a representative of the California Medical Association and often wrote about the migrant health program in various medical journals. Larsen wrote an addendum to Gregg's report, rebutting several of the U.S. Public Health Service officer's conclusions. Larsen countered Gregg's criticism about the high cost of hospitalization, for example, by noting that inadequate social supports, unstable family units, bad housing, and poor nutrition made early discharge from the hospital and clinic infirmaries problematic and inflated costs. Larsen also mentioned that 65 percent of initial visits occurred in clinics in 1941, compared with only 34 percent when Goldner first undertook his analysis. Larsen drew attention to this trend, a model he referred to as "primary contact," and suggested that with time it would restrain costs. Larsen conceded some of the organizational problems raised in Gregg's analysis, but he emphasized that five years of providing medical care to the nation's most indigent rural citizens through a partnership with the government had been an invaluable experience for practicing physicians and organized medicine.[96]

At first blush, it appears counterintuitive that a U.S. Public Health Service officer would criticize a program supported by his own agency and coordinated by some of his colleagues, while a representative of organized medicine would rally behind it. The paradox is more apparent than real. The California Medical Association was in fact more liberal than many if not most state medical societies (and it was certainly to the political left of the AMA). As medical director of the Agricultural Workers Health and Medical Association, Larsen was willing to work with the federal government on a medical care delivery program because the migrant health plan afforded strong financial benefits to participating doctors and it largely preserved physician autonomy and authority, even if it stretched conventional boundaries. Gregg, in contrast, criticized the association because it had not done enough to modify a style of medical practice that more progressive medical thinkers considered dated, inefficient, and costly. Gregg's support for salaried physicians and clinic-based care was therefore more in line with the views of social reformers, who aspired to make the practice of medicine more public.

Before World War II, protests against the migrant health plans on the grounds that they were a prelude to a compulsory government-controlled health care system were not often heard. However, some did voice this concern and others were troubled by the programs' ties to the U.S. Public Health Service, in particular the direct involvement of medical officers in providing medical care. From the AMA's perspective, this relationship was fraught with worrisome potential. This anxiety increased during the war, when the ties between the FSA and the U.S. Public Health Service expanded and both agencies became more active participants in the debate over national health programs and compulsory health insurance. Organized medicine also cast a wary eye at the Farm Security Administration when the agency initiated its boldest medical care program, the experimental health plans, in 1942. Within the agency, these experimental health plans were seen as an opportunity to explore more fully a range of organizational and financing strategies. Those interested in an expanded role for the federal government in health care also considered the experimental health plans a viable model for the delivery of comprehensive rural health services throughout the nation. From the perspective of organized medicine, however, the FSA was treading on increasingly dangerous ground.

CHAPTER FOUR

Modeling for National Health Insurance
The FSA Experimental Health Plans

I'm for the program 100 percent. Call it socialized medicine or what you will, it is helping a lot of farm families to get the best medical care they ever had, and that in the face of a wartime shortage of doctors. Every person ought to have access to the best medical care available, regardless of his ability to pay.
ARKANSAS PHYSICIAN, 1943

By the early 1940s, those directing the FSA medical care program had accumulated a vast fund of practical experience in providing health care services to low-income rural families. They were convinced that this experience could and should be used in planning postwar health policy for rural communities and not only for low-income farmers. From this point on, the FSA's involvement in medical care delivery was marked by bolder experimentation and an increasingly visible and controversial role in the national health policy debate. Beginning in 1942, the FSA organized, administered, and funded seven countywide health plans and one multicounty plan that extended membership eligibility to include rural families who were not rehabilitation clients and that explored different financing mechanisms. The experimental health plans, as they were called, marked a significant extension of the agency's medical care effort and indicated the conviction of agency leaders that these programs could be adapted to fit national health policy objectives.[1]

A number of considerations went into the decision to embark on a new health care initiative. The FSA's involvement in rural rehabilitation since 1937 had shown agency leaders that deficiencies in health care affected everyone living in rural communities. "The village drug store or the village general store, with a patent medicine counter, is the most characteristic rural health facility," wrote Fred Mott, who assumed the duties of acting chief medical officer in early 1942. "The gilded picture of the heroic country doctor is essentially a picture of poor medical care, with the lonely practitioner doing all in his power to make the best of the situation."[2] The truth of this observation was driven home in a dramatic fashion by the mobilization for war when the Selective Service rejected for health reasons over 50 percent of individuals employed in agriculture.[3]

Problems arising in the FSA's largest medical care effort, the medical care cooperative program, also contributed to the agency's interest in piloting a medical care delivery program with a broader membership base. For example, turnover in membership and variable enrollment continued to threaten the actuarial foundation of a growing number of medical care cooperatives. The FSA found that the families most likely to use services tended to maintain their memberships, while healthier families were more likely to let their membership lapse. These trends eroded a plan's membership base, shrank the pool of available funds, and ultimately decreased physicians' reimbursement and increased their dissatisfaction. For senior FSA staff, these problems underscored the fiscal weaknesses inherent in voluntary medical care programs, especially those aimed at low-income families.

Then there was the political environment in which the FSA was operating. In the years leading up to and including World War II, a conservative shift in national politics signaled growing disenchantment with the New Deal. As the nation's top priority became winning the war, Congress sharply curtailed federal spending for domestic programs. These factors reinvigorated the efforts of the agency's opponents to scale back FSA rural rehabilitation programs, including their medical plans. As the political and financial restraints on the FSA tightened, Health Services Branch personnel searched for solutions to address the fiscal and actuarial instability in the agency's medical care programs. By relaxing eligibility criteria and extending membership to more well-to-do farmers, it might be possible to place the agency's rural health care programs on more solid ground.

National health policy and national health politics also exerted potent influences on the development of the experimental health plans. The introduction of the 1943 Wagner-Murray-Dingell bill—which, if enacted, would have created a "cradle to grave" social insurance system, including comprehensive universal health insurance—reignited the smoldering debate over national health insurance. This bill generated guarded optimism inside the FSA that its rural health programs might become a model for a nationwide plan. In addition, the FSA's Health Services Branch was actively involved in health policy discussions within the executive branch during this period, and some of the more liberal members of Congress showed keen interest in the agency's health care delivery experiments in the context of postwar national health planning.

THE INTERBUREAU COMMITTEE
ON POST-WAR PROGRAMS

In 1941, the Secretary of Agriculture created the Interbureau Committee on Post Defense Planning and charged the group with developing "all possible phases of a health program for the entire farm population."[4] The committee was composed of representatives from the Agricultural Extension Service, the Bureau of Agricultural Economics (BAE), and the Farm Security Administration, all of which represented different political constituencies. After war was declared, this committee was renamed the Interbureau Committee to Coordinate Post-War Programs, and its mission broadened to include planning for postwar federal rural health policy. Despite the tension between the AES and the FSA, the working group did come to some resolution around the issue of rural health.

While the Interbureau committee did not formally propose a federally sponsored national rural health delivery system, it sketched out principles on which such a program would be based. In 1942, it submitted a report describing these principles and endorsing the demonstration projects under development by the FSA. The ideal plan described in the committee's report included general practitioner care, hospital coverage, drugs, basic dental care, and nursing services. Echoing the recommendations of prior health policy groups, and reflecting the influence of the FSA medical care programs, the committee recommended that any federal health program be centered around the general practitioner, who would also control access to specialist referrals, hospitalization, and medical technology. Funding for the program would come from annual dues paid by higher-earning families, with the federal government contributing toward membership fees for low-income families. Federal monies would be channeled to communities via a regional administrative hierarchy. The Interbureau Committee to Coordinate Post-War Programs also suggested linking county and state public health departments to rural health plans for a range of preventive services. In addition, it backed federal assistance for maternal and child health services, rural hospital construction, expansion of rural public health units, and increased support for medical research and education.[5]

Inside the FSA Health Services Branch, there was a sense of pride that the FSA had helped spawn a broader movement to improve rural health services. "The growth and popularity of the FSA medical care program," Chief Medical Officer Mott wrote in 1944, "naturally suggested patterns for the post-war planning of rural health services . . . It was, therefore, no surprise when the FSA was looked to for guidance in setting up extensive health programs in a selected number of rural counties."[6] Because of the competing interests within the Interbureau Committee, however, several of its recommendations fell short of the FSA's initial expectations. Farm Security Administration medical personnel were most disappointed in the Interbureau Committee's proposals dealing with physicians' services. To ensure the support of organized medicine and the farm bloc for its recommendations on hospital construction and improved public health facilities, the committee cautioned against "restricting

physicians services and . . . benefits in any way."[7] The committee supported
the FSA's experimental health initiative, but with two important limitations.
They insisted that the new program should be strictly voluntary and that the
Agricultural Extension Service would be given a dominant role in promoting
it among higher-income farmers. The Interbureau Committee's decision to rel-
egate the FSA's well-rehearsed field staff to a supporting role in promoting the
FSA's own experimental health plans was disheartening. Although FSA medical leaders understood the practical advantages of using the existing relationship between county extension agents and more affluent farmers to build
the program, they were equally cognizant of the AES's strong ties to the farm
bloc, whose opposition to the FSA was well known.

When FSA medical leaders first conceived of the experimental health program, they favored a flat dues structure of 6 percent of net income and opening the experimental health plans to all residents in the chosen county. An
open enrollment policy, they argued, would distribute the economic burden
of illness in any given community over a much broader patient base. This
would, in turn, stabilize federal subsidies and membership fees, and improve
physicians' reimbursement. Privately they also hoped that unrestricted membership would allow them to explore fiscal and organizational strategies that
might be of value in national health planning.

The idea of nonincome-restricted enrollment was quickly scrapped owing
to stiff opposition by county and state medical groups. By 1942, rural physicians were under less economic pressure than they had once been and held
even more tightly to the view that those individuals able to pay for medical care
in the traditional fee-for-service manner ought to do so. Consequently, the experimental health plans established income thresholds for eligibility and set
minimum and maximum membership dues based on net annual income. In
most plans, members paid between 4 and 6 percent of their annual income.
FSA medical leaders worried that limiting membership to families below a negotiated income threshold might cut into the actuarial strengths they hoped
to achieve in a nonincome-restricted health plan. They had little choice, however, but to agree to this policy and to use the plans to see how far the voluntary approach would take them. In practice, few families living in counties
where the plans operated were excluded because of income.

The communities that became sites for the experimental health plans were
selected using criteria suggested by the Interbureau Committee to Coordinate Post-War Programs. These included a community interest in improving
medical care, reasonably stable community income, sufficient medical and
hospital facilities, an active county agricultural planning committee, and
receptive local physicians. Geographic and agricultural diversity were additional considerations, as was a reasonably diverse socioeconomic and eth-

nic makeup. It was preferable, but not mandatory, that the community have a full-time public health unit. The presence of an FSA medical cooperative unit was a plus. A functioning medical care cooperative was a reasonable indica-tion that a community met some of the established criteria and that local physicians were sympathetic toward federal health initiatives.

Between July and November 1942, six rural health services and two "spe-cial area" demonstration projects were created with the financial and organi-zational support of the FSA. Plans were set up in Wheeler County and Cass County, Texas; Newton County, Mississippi; Hamilton County, Nebraska; Nevada County, Arkansas; and Walton County, Georgia. Of the two special demonstration projects, one was an ambitious six-county experimental health plan located in the bootheel of southeastern Missouri. The other was a plan for the largely Spanish-speaking population of Taos County, New Mexico. Like the agricultural workers health associations and many of the resettlement project cooperatives, these plans were administered locally through incorpo-rated entities. This approach gave them a strategically necessary distance from the federal government and enabled the FSA to claim that the experi-mental health plans held to the spirit of the AMA's policy on third-party in-volvement in financing medical care.

The various experimental health plans provided comparable benefits and were organized along similar lines. The guiding principles common to all in-cluded group prepayment, voluntary membership, local administration, and income-based annual dues. Families able to pay the annual membership dues, typically around $50, paid the full sum. The FSA subsidized all costs not cov-ered by membership dues and contributed to the dues paid by lower-income families.

The plans differed most in financing. Of the eight plans, six adopted mod-ified fee-for-service financing similar to the model used in most of the med-ical care cooperatives. Participating physicians billed a pooled fund generat-ed from annual membership dues and were paid on a monthly basis. If billings exceeded monthly allotments, doctors were paid on a prorated basis. The ex-perimental plans using fee-for-service financing were those in Nevada Coun-ty, Cass County, Hamilton County, Newton County, and Walton County, as well as the Southeastern Missouri Health Service. Two programs incorporat-ed nontraditional financing of medical services. The Wheeler County Rural Health Service operated on a capitated basis. In Taos County, the three-per-son medical staff received annual salaries and provided necessary medical care to plan members without submitting separate bills.

THE FORGOTTEN PEOPLE OF TAOS COUNTY:
A CASE STUDY OF ONE EXPERIMENTAL HEALTH PLAN

In the 1930s and 1940s, New Mexico was one of the nation's most rural states, and one of its poorest. The majority of families lived on land so harsh and infertile that Clay Cochran, an FSA advisor sent in 1942 to assess Taos County's potential as an experimental health county, classified the community's economic status as subsubsistence. "I have worked with the most exploited and disadvantaged groups in the state of Texas," noted Cochran. "I have never encountered a situation more tragic, or which demanded more immediate attention, than the problem of health in New Mexico."[8]

Like many rural states, New Mexico suffered terrible unemployment from the collapse of the agricultural economy that reached its climax during the Depression. At the time of Cochran's survey, fully 50 percent of all families in New Mexico were partially dependent on government assistance, and median farm income was a paltry $387 per year. Barely 15 percent of the populace had access to electricity and plumbing. Medical care was scarce and unaffordable. The state's infant mortality rate of 109 deaths per 1,000 live births was more than double the national average. According to Cochran's figures, the mortality profile for the state also deviated strongly from that of the nation as a whole. The rates of infectious diseases, such as tuberculosis, diarrhea, dysentery, influenza, and pneumonia; and preventable conditions, such as puerperal fever and accidents, were twice the national average, and each ranked in the top ten causes of death for the state. "These are rural agricultural people who are either the responsibility of the F.S.A.," wrote Cochran, "or they are really 'Forgotten People.'"[9]

Taos County was a persuasive site for other reasons. For one thing, the FSA could build on a community planning program initiated jointly by the University of New Mexico and the Carnegie Foundation. In addition, local physicians had participated in a medical care cooperative for 250 rural rehabilitation borrowers since 1940. Finally, the FSA was looking for an opportunity to pilot its experimental plans in communities with a significant number of minority residents. The Taos plan allowed it to develop a health care plan specifically for Spanish-speaking Americans. Similarly, Nevada County, Arkansas, was selected as a site partially because of the relatively large number of black families living there.[10]

In the summer of 1942, the Taos County medical society formally signed a memorandum of understanding with the FSA creating a new nonprofit association, the Taos County Cooperative Health Association. The FSA negotiated a sharply discounted fee schedule with the local medical and dental societies for care provided outside of the association's own clinics. Area hospitals offered reduced rates for Taos plan members as well. Local physicians, re-

Examination day at the clinic operated by the Taos County Cooperative Health Association. Peñasco, N.M., January 1943. John Collier.

spected community members, and farmers sat on the governing boards of each of the experimental health plans, although in Taos County, FSA representatives were a majority on the board.

The FSA provided funds to build a large clinic in Taos and three satellite clinics in the neighboring communities of Peñasco, Cenna, and Questa. Three full-time salaried nurses and one nursing supervisor staffed these facilities. A full-time salaried medical director, two full-time physicians, two part-time surgeons, and one salaried dentist provided medical, surgical, and dental services. The medical director received an annual salary of $5,200 per year, while the rest of the physician staff were paid $3,000 annually. Nurses received $1,800 per year. During World War II, when health personnel were in short supply, the association also paid $900 per year plus living expenses to two Mex-

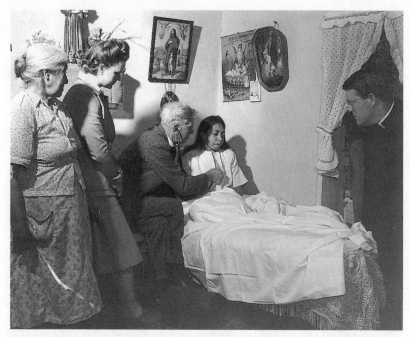

Doctor Onstine, of the clinic operated by the Taos County Cooperative Health Association, and Father Smith, the parish priest, at the bedside of a tuberculosis patient. Questa, N.M., January 1943. John Collier.

ican nationals. Both doctors were recent medical school graduates who used their positions in Taos to fulfill mandatory national social service obligations.[11]

The services available in the Taos County Cooperative Health Association were comprehensive and typical of the experimental health program as a whole. Members were eligible for general practitioner services, including office, home, and hospital visits. Obstetrical services, fracture setting, and major and minor operative procedures were standard benefits as well. The Taos plan maintained the preventive emphasis so important in the migrant health programs, including childhood immunization, nutritional counseling, and maternal and infant health programs.[12]

In an effort to control costs, the experimental health plans in Taos and elsewhere placed more restrictions on surgical procedures, specialty consultation, and hospitalization than did most of the other FSA medical care plans. Routine hospital costs, such as anesthesia, radiology, laboratory testing, operating room overhead, and nursing services were covered, but the program limited hospital stays to fourteen or fifteen days. Longer stays required authorization either from the association's medical staff or from local FSA supervisors. Preauthorization from either the attending physician or, less commonly, from a consulting doctor, was also required for patients to use medical services not available within the county.

Even within this more restrictive atmosphere, however, the FSA maintained its characteristic flexibility and pragmatism. FSA leaders acknowledged that local concerns required local solutions and required the input of participating physicians. In a 1942 letter, the FSA's chief medical officer noted that "although the experimental health proposal has been outlined in fairly specific terms, it is recognized that due allowance must be made for flexibility . . . provided that there is no sacrifice of the essential principles."[13] Taos County Cooperative Health Association members, for example, were offered ambulance service and prescription eyeglasses, and—with preapproval by staff physicians—they had access to specialized surgical and ophthalmologic services at Santa Fe medical institutions.

Because referrals to specialists outside the patient's immediate area were not common medical practice in many rural communities, the FSA saw such policies as an improvement in the quality of health care potentially available to participating families. In practice, however, there were few such out-of-county referrals. Between October 1942 and September 1943, there were approximately 7,500 clinic visits, 2,700 days spent by members in hospitals, and 1,400 dental visits in the Taos plan. During the same period, there were only 10 referrals to outside specialists and only 172 referrals to ophthalmologists. Rural physicians were accustomed to being the sole health-care providers for their patients, and they were not eager to relinquish control over their patients to other doctors. FSA medical leaders acknowledged the divergence between policy and practice, but they felt the problem did not affect the overall quality of care available to Taos plan members. For this reason, and for obvious tactical reasons, they deferred to local physicians' preferences.[14]

The dental services available in the experimental health plans were more extensive than the emergency care typically provided in other FSA medical programs. Taos plan members, for example, were also entitled to such preventive measures as cleaning, amalgams, gingival treatments, X-rays, and routine checkups. Initially, drugs listed in the *National Formulary* were paid for in the experimental health plans, but rapidly rising costs and opposition by local pharmacists eventually caused the discontinuation of this coverage.[15]

Enrollment in the Taos County Cooperative Health Association, as in all experimental health plans, was voluntary. Residents with an annual income below $1,200 were eligible to join. This was substantially lower than the $1,800 income limit set in other experimental health plans, but given the poverty of the county, the lower ceiling excluded few residents. In the first year of the Taos plan, over a thousand families enrolled. The FSA took this as an indication that the plan was favorably received by the community. From 1942 to 1944, approximately 40 percent of the county's entire population (representing more than two-thirds of all farm families) joined the association. Taoseños made up the largest group of plan membership as a percentage of population, with relatively fewer families joining from Questa, Cenna, and Peñasco.[16]

MEMBERSHIP AND COSTS IN THE EXPERIMENTAL HEALTH PLANS

The experimental health plans required a substantial federal subsidy to keep dues low enough to encourage membership for those who were not FSA borrowers and to keep services comprehensive. The Farm Security Administration covered 80 percent of the cost of the experimental health plans in their first year of operation. Most of the remainder of the funding came from local contributions of land, supplies, and labor (10 percent) or from membership dues not subsidized by the FSA (7 percent). Nonfarm families and individuals occasionally joined the experimental health plans by paying dues based on income, but this income amounted to just over 1 percent of the total program expenses.[17]

Funding costs for the experimental health plans plateaued in 1943. This was welcome news for the FSA given Congress's tight-fisted policies during the war and the agency's declining appropriations. As an example, the FSA's subsidy to the Taos County Cooperative Health Association dropped to 60 percent of costs in its second year of operation. The FSA attributed this to the leveling off in utilization by plan members, shorter hospital stays, and a strong rally in farm prices that boosted dues collections and encouraged additional families to join the plan. However, while the experimental health programs were on a sounder financial basis by 1943, they never came close to being self-supporting. In its final year of operation, the FSA still provided 50 percent of the program's funding.[18]

Nationally, membership in the experimental health plans climbed to nearly 35,000 individuals over the course of the first year, and peaked the following year with approximately 42,000 individuals enrolled. After 1943, membership in the experimental health plans declined steeply in a trend that paralleled that seen in the FSA's medical care cooperatives. Enrollment fell to 32,000 persons in 1944 and to fewer than 24,000 by the end of 1945.

As a percentage of the local farm population, experimental health plan membership ranged from a low of 32 percent in Wheeler County, Texas, to a high of 74 percent in Taos County, New Mexico, in the 1942–43 fiscal year. For all plans combined, 50 percent of eligible persons actually joined the program. Moreover, the experimental health program experienced the same problems in recruiting and maintaining members as had the medical care cooperative program. Approximately 50 percent of plan members maintained their membership from year to year. In Wheeler County, the situation was even more dismal. According to the BAE, less than 8 percent of families maintained continuous membership over a three-year period.[19] The FSA medical leaders had anticipated greater rates of participation and more stability in terms of enrollment in the experimental health plans than was the case. This contributed to the actuarial difficulties the program experienced, and it would later figure in the decision of senior Health Services staff to support national health legislation.[20]

When queried about their reasons for letting their memberships lapse, families commonly cited high annual dues, or the fact that they did not receive the services for which they paid, or that they wanted. Most members knew what services were covered, but they were unaware that the FSA was underwriting over 80 percent of the program's actual costs. Lack of drug coverage was a frequent complaint as well. "What's the use of going to a doctor," a Nevada County plan member complained, "if you don't have the money to buy the medicine he tells you to get?"[21]

The anemic membership figures for more affluent farmers were especially disappointing to FSA medical leaders. There were several reasons why the experimental health plans failed to enroll as many higher-earning families as the FSA had hoped. For one thing, there was a popular perception of the program as a plan for the indigent. Many upper-income farmers simply viewed the new medical care plan as a charity program for disadvantaged farmers, an understandable perception given the FSA's primary legislative mandate. In some plans, this perception was played on by local physicians, who actively discouraged better-off families from joining the experimental health plan, or suggested that they not renew their memberships once their economic status improved. In addition, not all families were persuaded that advance payment for medical care made good economic sense, an indication that the concept of group health insurance had still not been accepted by many rural families.[22] Finally, FSA leaders suspected that county extension service agents failed to enthusiastically promote the experimental health plans among higher-income families.

AN INCREASED FEELING OF SECURITY: THE IMPACT OF THE EXPERIMENTAL HEALTH PLANS

Health statistics for rural communities in the 1930s and 1940s vary in accuracy and completeness. However, there are data that indicate that the majority of participating families perceived the health care they received in the experimental plans to be equal to or an improvement over what they had been used to. When surveyed by the BAE, 98 percent of all experimental health plan members stated that the program provided better or at least equal med-

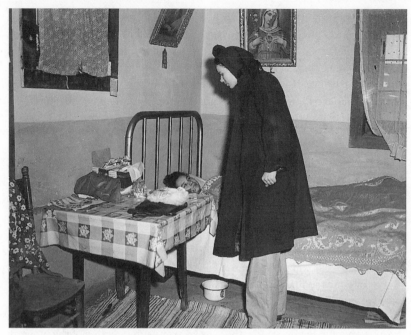

Marjorie Muller, nurse from the clinic operated by the Taos County Cooperative Health Association calling on an obstetrical case who is to go to the hospital for delivery. Peñasco, N.M., January 1943. John Collier.

ical care than they had previously experienced. In some counties, the improvement was more noticeable to plan members. In Taos County, for example, 80 percent of plan members stated that they received better medical care under the new program. In the other experimental health plans combined, however, only 40 percent felt they received better care under the new arrangement.[23]

Hospital-based, physician-attended births have historically been considered an indicator of access to quality medical care, and the FSA pointed proudly to the record of the experimental health plans in this area. The plans built on the policies adopted by the FSA from the beginning for what the agency considered to be an important improvement in the health care of pregnant women and babies. The Taos plan, for instance, established policies that required a minimum number of prenatal and postnatal physician visits, hospitalization for delivery, and immunization of infants. For the majority of Taos County women, this was the first time their babies had been delivered by a physician. A 1944 BAE report on the Taos plan commented that "more mothers are now receiving prenatal and postnatal care, and better medical care and more hospitalization at childbirth."[24]

Nevada County, Arkansas, offers more dramatic evidence of the impact of the experimental health plans on obstetrical services. Before the creation of the Nevada County Rural Health Service, 90 percent of all babies born to white mothers in the community were physician attended, but only 10 percent of these births occurred in the hospital. African American women in Nevada County delivered at home 99 percent of the time and were rarely attended by a physician. Following the introduction of the experimental health plan, obstetrical practice in the county was transformed. Fifty percent of all white babies and nearly 25 percent of black infants were delivered in the hospital, and physicians now delivered three-quarters of all black babies.[25] Whether or not infant and maternal mortality rates improved as a result of these policies was not documented at the time by the FSA. However, local doctors, rural families, and the FSA shared a belief that this was the case.

Another criterion used by the FSA for evaluating the success of the experimental health plans was utilization. Volume of services was one of the most comprehensive and detailed items of statistical information gathered by the FSA and by other agencies, such as the BAE. In the first year of the experimental health program's operation, demand for services exceeded the FSA's original projections, forcing the agency to replenish exhausted cash reserves halfway through the fiscal year. In Taos County, 94 percent of plan members received some medical care during the first year of operation. On average, experimental health plan members saw a physician 2.6 times a year, a figure that equaled that for the general population and was more than double what rural families from the same income group received in communities not served by an FSA plan. FSA medical leaders took less solace in these figures than one might expect. For example, the utilization figures were well below the 6.6 visits per year that they cited as being necessary to achieve the level of medical care rural families needed.[26] Members in the southeastern Missouri experimental health plan received three times as much medical care as they had before the plan went into effect, raising them to a level of utilization that was 75 percent higher than that of the typical rural dweller and 50 percent higher than that of the population at large.[27]

Membership in the experimental health plans decreased barriers that had prevented families from seeking medical care and gave members a sense of security. According to the BAE, Taos County Cooperative Health Association members went "to a doctor oftener and earlier, and are making more use of hospitalization facilities."[28] A 1943 survey found that Taos County Cooperative Health Association members overwhelmingly favored the medical care plan. Fewer than 3 percent of those surveyed by the BAE specifically opposed federal aid for rural health services, although about 20 percent were undecided.[29] As one formerly prosperous farmer in Wheeler County, Texas, noted: "Government should stay in this health program permanently and extend it to

TABLE 4.1 RATES OF UTILIZATION BY TYPE OF SERVICE IN THE
EXPERIMENTAL HEALTH PLANS AND IN THE GENERAL POPULATION,
1942–1943[a]

Type of service	Experimental health plans	General rural population
Physicians' services	1,435	526
Office calls	2,663	1,040
House calls	211	530
Total	2,874	1,570
Surgical cases	71.7	47.7
Hospital service		
Admissions	112	42
Days hospitalized	423	505
Dental cases	300	159

Source: Frederick D. Mott, *Health Provisions in a Social Security Program for Farm People, with Special Reference to the Experience of the Health Programs of the Farm Security Administration* (Washington, D.C.: U.S. Department of Agriculture, Farm Security Administration, 1944), 3, file 31-2, p. 19. Frederick Dodge Mott Papers, Canadian National Archives, Ottawa.
[a] Numbers are per thousand persons per year.

a wider area. Some people don't like to have anything to do with an organization if the Government's in it, but with me it's the other way. That is, if the Government's backing a program, that's a guaranty [*sic*] you won't get swindled."[30]

Rates of hospitalization, surgeries, and obstetrical cases were above national averages in each experimental health plan. Interestingly, however, compared with other rural groups over a one-year period, experimental health plan members had shorter hospital stays: 3.8 days versus 12 for rural communities as a whole (Table 4.1). This may have been due to the experimental health plans' administrative restrictions, or their more explicit cost consciousness. Alternatively, high rates of hospitalization and shortened duration of hospitalization may indicate that the threshold for hospitalization was low enough that less acutely ill patients were being admitted.[31]

The increased use of medical services emphasizes the point that rural citizens had significant medical problems and desired more medical care. Reasonably high rates of membership attested to the fact that many were open to proposals that would address their needs. As the wife of one farmer in Wheeler County commented:

> We farm, and help produce the stuff the Nation needs. So, if we don't make enough to buy the health attention we need, or to keep the doctors' incomes high enough so they can have all the equipment and give all the service they ought, and if the Government steps in and helps to bear or equalize the cost, I

don't feel that they're giving us something for nothing. I can join the association and take the assistance the Government gives and still feel we don't owe the Government anything and that they don't owe us anything. That's only doing right by people.[32]

The FSA expected that higher-income families would be healthier and require less medical care than those in lower socioeconomic groups. This was not the case. Better-off families, nearly all of whom were white, used more services in the experimental health plans than poorer and minority members (Table 4.2). The reasons for this are speculative, since the historical record is generally silent on this issue. As noted earlier, the communities selected for experimental health plans were located in counties where an FSA medical care unit was already in operation. Possibly the established medical care cooperatives ameliorated pent-up demand for medical services by the most medically needy farmers, thereby inflating the utilization rate among families who were not covered but who had significant medical care needs of their own. When analyzed by month of operation, utilization decreased over time, suggesting that the initial burst of physician visits may have been due to farmers getting medical care they had postponed for financial reasons. Well-to-do members may also have wanted to make certain they got their money's worth, or perhaps they were better able to get medical care from participating physicians.

There are clues suggesting that some doctors preferentially treated higher-income families in the experimental health plans and that they were less inclined to provide services to minority or poorer families. Preferential treatment

TABLE 4.2 ANNUAL PHYSICIAN VISITS IN TWO EXPERIMENTAL HEALTH PLANS, BY INCOME AND RACE, 1944

	Number of visits	
	Cass County	Nevada County
Income level per family		
<$250	4.5	3.0
<$500	4.2	2.8
≥$500	5.4	5.1
Race		
White	5.0	4.0
Black	3.1	1.3

Source: Carl C. Taylor, T. Wilson Longmore, and Douglas Ensminger, *The Experimental Health Program of the United States Department of Agriculture*, prepared for the Senate Committee on Education and Labor, Subcommittee on Wartime Health and Education, 79th Cong., 2d sess. (Washington, D.C.: GPO, 1946), p. 35.

of higher-income and nonminority families by physicians was possibly influenced by class and racial considerations, but there were clear economic advantages for doctors as well. Doctors may have been more solicitous of higher-earning families because they would be the first to reenter the traditional fee-for-service market when the experimental plans ended, or when a family's annual earnings moved them over the income threshold. [33]

Historically, the lowering of access barriers to medical care has invariably resulted in greater use of health care services. This phenomenon has been noted following the initiation of other government-sponsored health insurance programs in the United States, such as Medicare and Medicaid. What is noteworthy here is that it was observed at a time before the generalist-to-specialist ratio had become dramatically imbalanced, before hospitalization became common in rural communities, before diagnostic technologies became plentiful and widely available, and just as third-party health insurance was becoming a noticeable wave in a sea of fee-for-service practice. In short, this phenomenon occurred before many of the factors that are credited with driving the explosion of health care utilization and rising costs in the last quarter of the century had come into being.[34]

The experimental health plans had much in common with the FSA's other medical care programs, and the agency drew on its earlier efforts in developing what turned out to be its final rural health initiative. The use of general practitioners or nurses at the center of clinical decision-making, administrative restrictions on specialist referral and hospitalization, the focus on prevention and health education, and the promulgation of maternity and postnatal practice guidelines were familiar features. Experimental health plan participants—farmers, physicians, and the FSA alike—believed these policies represented an improvement in the quality of medical care provided to plan members. What made such policies more significant in the experimental health plans is that FSA medical leaders (and others) saw the new program as a pattern for nationwide application. Consequently, they were more carefully planned and more restrictive than in, for example, the medical care cooperatives.

Underlying the FSA experimental health plans was a deeper aim that the agency very rarely called attention to: reshaping the practice of rural health care. In a confidential 1942 letter, the FSA's chief medical officer acknowledged that the FSA sought to transform how doctors practiced medicine in rural communities. "If we are to make a fundamental approach toward solving the sort of problem presented," wrote Fred Mott, "we must not be content with perpetuating the existing pattern of medical practice."[35] Two years later, Mott again hinted at the FSA's sub rosa agenda in a speech delivered to the American Public Health Association. In his talk, the chief medical officer contrasted the medical care cooperative program with the experimental health plans. In the former, noted Mott, "the pattern of providing medical care was not altered from the usual pattern of private medicine." Mott then used Taos County as an example of an experimental health program where "it was possible to establish a system of medical services quite different from the conventional pattern of private, competitive practice."[36] Certainly, in the area of maternal and child health, and to some extent in regard to control over specialty and hospital referral, the FSA moved beyond fiscal and administrative roles in the experimental health plans.

PHYSICIANS' RESPONSES TO THE EXPERIMENTAL HEALTH PLANS

Physicians' attitudes toward the experimental health plans were predictably mixed and evolved as circumstances at the local or national level changed. The same economic and philosophical concerns raised by the medical profession in the FSA's other medical programs also factored into physicians' views on the experimental health plans. There were complaints about low fee schedules and reimbursement. General practitioners in the southeastern Missouri experimental health plan were unhappy that they received a lower percentage of billings than did surgeons, hospitals, or dentists participating in the program. Eligibility criteria, too, remained a sore point, as did rules about referrals and hospitalization. Some doctors were pleased with policies that reinforced their central role as providers. Other general practitioners, specialists, and the leaders of organized medical groups sometimes complained that such restrictions were examples of the bureaucracy they warned would come from federalized medicine.[37]

There was a continuing distrust of medical care programs based on anything but fee-for-service financing. Capitation was particularly foreign to the great majority of physicians and they distrusted it as a form of contract practice.[38] Critics feared that capitation might encourage doctors to restrict needed medical care and services, but supporters felt there were safeguards that would prevent this. "Under the capitation plan," a member of the Wheeler County Rural Health Service was quoted as saying, "it also might seem to the doctor's advantage to discharge patients a little early or go slow on the expensive medications and serums if hospitalization and drugs were paid for by a fixed annual fee per family. But that sort of abuse wouldn't happen much in local neighborhoods where everybody knows what happens."[39] The Farm Security Administration doubted that abuse occurred in more than a few instances and suggested that professional ethics and doctors' neighborly relationship with their communities would adequately protect against it.

In those counties in which the plans operated, local practitioners were by and large supportive, and only one county medical society involved in the pro-

TABLE 4.3 MONIES ALLOCATED BY EXPERIMENTAL HEALTH PLANS,
1942–1943 AND 1943–1944

| Type of service | Dollars allocated in membership fees[a] | |
	1942–43	1943–44
General practitioner	$16–22	$16–19
Surgeon-specialists	$6	$6–7
Hospitalization	$8–12	$9–12
Drugs	$5–7	$0–6
Nursing	$0–3	$0
Dental	$6–7	$0–7
Administration	$2–4	$3–5
Total	$50–57	$41–50

Source: Carl C. Taylor, T. Wilson Longmore, and Douglas Ensminger, *The Experimental Health Program of the United States Department of Agriculture*, prepared for the Senate Committee on Education and Labor, Subcommittee on Wartime Health and Education, 79th Cong., 2d sess. (Washington, D.C.: GPO, 1946), p. 35.
[a] Taos County data are not included, although annual costs per capita were $35.59 (1942–43) and $51.48 (1943–44).

gram discontinued its participation before federal funding was formally withdrawn in late 1946. Despite the complaints of some, physician reimbursement remained a strong point in the plans' favor, even though financial pressure on rural physicians had eased by this time (Table 4.3). In each of the five experimental health plans that used modified fee-for-service reimbursement, physicians received an average of 90 percent of billings or better during the plans' first two years. This figure was much higher than what rural physicians traditionally received from their practices, and this was duly noted and approved by local professional organizations.

Just as important, nonparticipating doctors and county medical groups did not mount any sustained opposition to the experimental health plans. An analysis of the plans by the Bureau of Agricultural Economics found that nonparticipating medical societies and physicians were either ambivalent or cautiously supportive of the program. The majority of those surveyed took a pragmatic position. Most were unwilling to pass judgment on the decision of their peers to participate in the experimental health plans, while others were indifferent, or simply were not that concerned about the program's implications.[40]

This support was nurtured by the FSA's continued reliance on the politically expedient policies that guided its earlier efforts. Local physicians and county medical societies were included in the development and operation of the experimental health plans and the plans continued to include income-based eligibility, voluntary participation, and free choice of physician. So long as the agency maintained these policies, its experimentation with financing mechanisms, its attempts to broaden membership beyond the indigent, and its expanded administrative control over medical practice seem to have been within boundaries that most participating physicians and county medical societies found bearable. In contrast, no state medical society except the Missouri State Medical Association gave its blessing to any of the experimental health plans.[41]

The American Medical Association's response to the experimental health plans was guarded. The AMA repeated its view that federal health initiatives were not justified, especially given that the social and economic landscape had improved for most Americans. The experimental health program "deserves the most careful consideration," commented the *Journal of the American Medical Association*, "since it involved several subjects that have been previously considered but under somewhat different circumstances."[42] Although the AMA did not reject the idea or work actively against the FSA in setting up the experimental health plans, at the 1942 house of delegates meeting, the organization adopted a resolution withholding its endorsement of the FSA plans. Consistent with its earlier approach to the FSA, the AMA instructed the agency to pursue the matter with county or state medical societies.

It is telling that the most potent physicians' organization in the country did not take a more forceful stand against the Farm Security Administration's experimental health plans. There are several reasons for this, ranging from the program's economic and professional reinforcement of allopathic physicians to organized medicine's concern with proposals that appeared more menacing. "There are almost parallel movements throughout at least the English-speaking world in the direction of greater intervention of government in medical care . . . ," one AMA official told his colleagues. "It is doubtful if head on opposition can greatly affect this trend."[43]

In exchange for their cooperation, many physicians and medical societies believed that they had increased their leverage with the federal government in the debate over national health politics. Lorin Kerr remembered his astonishment when he heard of a surgeon in Washington State who was "doing appendectomies for 75 bucks" for FSA families while other surgeons charged $125. When the FSA's district medical officer, Allen Koplin, arranged a meeting between Kerr and the surgeon in Yakima, Kerr found out why: "Doctor, I've been watching the legislation, what's coming along," the surgeon told Kerr. "Eventually, there's going to be national health insurance, like it or not . . . I have a hunch that you and Koplin are going to be involved with it and I just don't want you to forget that I was willing to work with you!"[44] A 1943 *Journal of the American Medical Association* made this point more explicitly when it told its readers that the FSA had "adopted a policy of close cooperation with state and county medical societies, and is pledged not to introduce any plan against the opposition of these groups."[45]

On the other hand, some in the medical community considered the experimental health plans a blatant step toward government-controlled medicine. In southeastern Missouri, the Dunkin County Medical Society refused to participate in the seven-county plan proposed by the FSA because in their view it represented unfair federal competition for the private practitioner. In his Taos report, Clay Cochran had also noted that several local physicians expressed concern that the experimental health plan might lead to government-controlled medicine. There was little about the experimental health plan in Nevada County that satisfied at least one Arkansas doctor: "If folks want to know what I think of the Nevada County Rural Health Association, you might just tell them that the thing ain't worth a———. I'm opposed to the program for the following reasons: 1. It is a step toward federalized medicine. 2. I have nothing to say in setting [sic] amount of fees. 3. The program is abused by members. 4. The program is abused by doctors. 5. Doctors' bills are scaled down and have not been paid 100 percent."[46] While critics of the FSA medical programs had always warned that they were opening the door to further government involvement in medical care, the stridency of the complaints leveled at the experimental health plans clearly reflected the intensified debate on national health issues during the war.

Those who raised the specter of "state medicine" had a point. While senior FSA medical leaders did not discuss it with rural doctors and medical societies, they were sanguine that the experimental health plans were the start of a broader federal rural health initiative in the context of national health planning. The work of the Interbureau Committee to Coordinate Post-War Programs was just an indication of renewed government commitment. In a 1943 letter, Mott informed his staff: "This should be considered confidential but at the national level we are trying . . . to make it possible for you to deal directly with the State Extension Services in developing programs for the general rural population."[47]

The experimental health plans operated for four years before the FSA itself was terminated by Congress. Although FSA leaders hoped the plans would survive the agency, they were unable to become self-sustaining. Still, for a period of some four years the experimental health plans provided medical care to a significant number of families in selected rural communities and gave agency leaders, farmers, and physicians valuable experience with a comprehensive group prepayment medical care program that nominally at least included nonindigent farm families. In the end, they were more notable for what they demonstrated was possible under the auspices of the federal government than membership figures or length of operation might imply.

When FSA leaders came out publicly in support of national health insurance legislation in 1943, the collective reaction by medical societies at all levels was loud, vitriolic, and sufficient to turn many formerly supportive physicians against the FSA. The agency's decision was a rude shock to some doctors, and a vindication for others. Once the agency publicly joined the battle over national health insurance, the fate of the FSA medical programs was sealed. Unfortunately for the FSA, this occurred just as the agency was at its most fragile—financially strapped and reeling in the face of a withering assault by conservative politicians and agricultural groups.

CHAPTER FIVE

The FSA Goes to War

These agencies in the government that are seeking to bring about
fundamental changes in the nature of medical practice are aiming
particularly at the farm groups . . . This plan has been sold to the medical
profession of this country.
MORRIS FISHBEIN, M.D., JOURNAL OF THE AMERICAN
MEDICAL ASSOCIATION, 1940

The disintegration of the Farm Security Administration and its medical programs began in earnest after 1942. Some of the reasons behind this demise were not specific to the FSA but broadly affected all New Deal social welfare programs. In this category fell Roosevelt's preoccupation with international affairs, Congress's growing conservatism, and waning public support for the New Deal. Other factors directly affected the FSA and its medical programs. These included the improved economic condition of doctors; intensified efforts on the part of the farm bloc to cripple the FSA; the growth of private health insurance plans; the shift of rural populations, including physicians and other health personnel, to the war effort; and the politics of national health insurance.

World War II catalyzed a demographic transition with roots traceable to the post–Civil War period. Military service and the promise of a well-paying job in war industries hungry for labor lured millions of rural Americans from their homes. Most never returned. Because their socioeconomic status was the most precarious, low-income farmers, migrant workers, and southern blacks constituted a disproportionate fraction of this population. Rural physicians and nurses volunteered and were drafted for military duty in greater numbers than their urban colleagues, exacerbating an already chronic shortage. By 1944, more than eighty rural counties had no practicing allopathic doctors, and an additional sixty-one had only a single active physician. A significant number of rural physicians dusted off their black bags and came out of retirement in order to help alleviate the crisis caused by the loss of so many younger practitioners. The average age of rural medical practitioners was historically higher than the national average; now it rose even further.[1] These developments had obvious consequences for an agency whose primary mandate was the rehabilitation of the nation's poorest farmers and whose medical programs depended on the cooperation of local physicians.

In their 1948 textbook, *Rural Health and Medical Care,* Mott and Roemer attribute most of the problems experienced by the FSA to changes wrought by the war, which were largely out of the control of the FSA. To the degree that subsequent historians have commented on the FSA medical programs, this explanation has been generally accepted. However, the war had a positive as well as a negative impact on the FSA and its medical care programs. On one hand, opponents of the FSA successfully pressed Congress to trim the agency's programs, claiming that the need to concentrate the nation's resources on the war mandated a leaner and more efficient government. As the nation's military involvement deepened, this argument gained credibility in a Congress that was eager to restrain domestic social programs that many felt had grown bloated, redundant, and inefficient at the trough of the New Deal. On the other hand, the importance of food production to the war effort led to unprecedented federal control over the farm economy and, in turn, to a significant expansion of the FSA health programs for farm workers. FSA leaders also emphasized the role that small producers could have on wartime food production, and the agency's effort to capitalize on the "small farmer issue" met with moderate success.[2] While the cumulative impact of these forces on the vitality of the FSA and its medical care programs was largely harmful, this chapter illustrates how the war years offered both risk and opportunity for the FSA and its medical leaders.

NATIONAL POLITICS AND THE FSA, 1940–1943

The New Deal's popularity in the polling booth began to wane in the 1938 midterm elections. Although Democrats held onto a slim majority in both legislative chambers, they lost 81 House seats and 8 Senate seats to the Republicans. Many returning Democrats were conservatives whose reservations about the president's social programs had prompted Roosevelt to campaign openly against them in their reelection bids. This rightward political trend continued in the 1940 and the 1942 elections. By the time the United States entered the war, the democratic coalition that built the New Deal was moribund.[3]

The prevailing political winds brought a period of escalating turmoil for the FSA. Until 1941, skillful political maneuvering by the FSA's top administrators, in particular Will Alexander and his successor, C. Benham ("Beanie") Baldwin, parried the most damaging attacks on the FSA. After 1941, these attacks sharpened. With the backing of the Grange and the Farm Bureau, anti–New Deal politicians in Washington steadily chipped away at the agency's programs. The farm bloc encouraged Congress and the president to place the FSA under the jurisdiction of the Agricultural Extension Service as part of a larger restructuring of the Department of Agriculture. The alliance also pressured Congress to investigate the FSA's activities and to curtail both its mission and its budget.[4] These efforts were not always successful. For example, the FSA was not placed under the jurisdiction of the AES. Over time, however, the offensive had its intended effect. Venomous appropriations battles, declining budgets, hostile congressional inquiries, and highly personal attacks on agency leaders forced the FSA into an increasingly defensive posture throughout the war.

Beginning in late 1942, a series of bureaucratic reorganizations, congressional actions, and high-level politicking dealt the FSA a succession of defeats. The first change came in December 1942. As part of a partial reorganization in the Department of Agriculture, the FSA was moved into the department's Food Production Administration. The latter in turn was placed under the War Food Administration (WFA) in March 1943. Throughout these initial bureaucratic reassignments, the FSA maintained control over its rehabilitation programs, budget, and staff. This changed a few months later. In June, the FSA suffered one of its most significant programmatic losses when Congress voted to move the migrant labor programs out of the FSA and place them directly under the control of the WFA. Although this decision was presented as a matter of wartime efficiency, it occurred without consultation with the FSA leadership. This was a serious setback to the FSA's influence and self-esteem, since the agency had played a major role in migrant welfare issues for the better part of a decade.

These reorganizations took place in the context of a series of emotionally charged congressional inquiries into the FSA, and appropriations hearings marked by caustic attacks on the agency and its staff. The first of two congressional investigations began in 1941 under the direction of the Joint Committee on Reduction of Non-essential Federal Expenditures—a name specifically chosen to emphasize Congress's aim of curbing federal spending in order to better prosecute the war. Chaired by one of the FSA's most indefatigable opponents, Senator Harry Byrd of West Virginia, the committee conducted a thinly veiled inquisition of the New Deal in which the FSA came under especially intense scrutiny. FSA leaders rallied their supporters in Congress, in the Roosevelt administration, and from around the country to rebut the charges made against it. The FSA survived the Byrd committee hearings, bruised but still intact.

Congress pared nearly all social welfare programs during the war, but it wielded an ax when it came to the FSA's budget. In 1941, the FSA's budget of $289 million supported a staff of some 19,000 and funded approximately 80,000 standard rehabilitation loans and 8,000 tenant purchase loans. Although the FSA's budget never allowed it to reach more than a fraction of po-

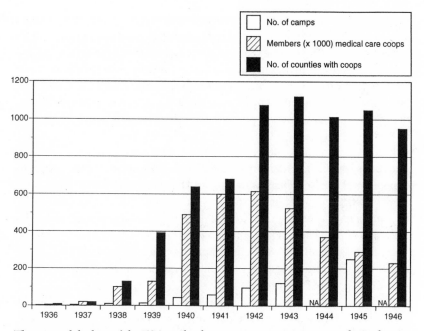

The rise and decline of the FSA medical care programs, 1935–1946: the evolution of the FSA medical care cooperatives and the migrant camp program from 1937 through the end of World War II. Although the peak year for the agency was 1942, the number of counties in 1943 was higher only because of wartime consolidation. Annual Reports of the Chief Medical Officer, 1935–46.

tentially eligible rural families, the agency was involved at some level with 800,000 of the estimated 7 million farm families whose annual earnings fell below $250. In 1942, Congress slashed the FSA's appropriation by a third, and by 1943 it stood at just $111 million. Between its shrinking budget and the wartime draft, the FSA lost experienced staff at all levels. Full-time personnel fell to 14,862 in 1943 and to 8,742 by 1945. For an agency whose programs were dependent on field staff, these losses had a devastating effect. With the exception of the migrant health programs, the FSA medical care programs saw parallel declines in terms of staffing and funding. Medical care unit terminations accelerated and membership in the medical care plans declined by more than 50 percent from 1942 to 1945.[5]

The process of cutting the FSA budget was marked by a bitterness experienced by few other federal agencies. Partisan politics had as much to do with the agency's escalating troubles as the need to prosecute the war efficiently. The FSA had had enemies all along; now they used World War II as a convenient rationale for attacking an agency whose programs and philosophy they had long despised. During the 1942 budget debate, for example, Senator Byrd castigated the FSA for administrative wastefulness and argued that the agency's programs would be managed more efficiently by the Agricultural Extension Service. In the same year, there was also a damaging political flap over the FSA's policy of including the payment of poll taxes in its rehabilitation loans to African American farmers in the South. The FSA policy, which enabled previously disenfranchised African Americans to vote, drew the condemnation of many southern politicians for subverting the established social order.[6]

Equally detrimental to the FSA were the increasingly strident denunciations of its leaders. FSA Administrator Will Alexander was a favorite target for opponents of the New Deal, as Rex Tugwell had been before him. By the summer of 1940, Alexander felt that the European conflict was sapping Roosevelt's commitment to the New Deal's social programs. Alexander had also tired of the barbs directed at him and at the FSA, and in June of that year he left the FSA for the Julius Rosenwald Fund, a private philanthropy interested in voluntary hospital insurance programs and in cooperative approaches to social problems.

Alexander's successor, Beanie Baldwin, was a veteran of the U.S. Department of Agriculture and a Tugwell disciple. Over the next three years, Baldwin proved to be a capable and creative administrator, as well as an effective advocate for the FSA's rural rehabilitation programs. He guided the FSA through some of its most difficult years before he too came under pressure to resign. During the 1942 appropriations hearings, Senator Kenneth McKellar of Tennessee accused Baldwin of being "not far from a communist" and claimed that the agency was promoting socialized medicine.[7] FSA administrators in state and regional offices were subjected to similar innuendo and overt character assassination. At a 1943 congressional hearing on the migrant health programs, one member of the California congressional delegation called the state's regional director, Jonathan Garst, a "Communist." When Beanie Baldwin defended Garst—in the process abandoning his characteristic Virginian equanimity by calling Garst's accuser a "choice but unspecified epithet"—the congressman peeled off his jacket and challenged Baldwin to fisticuffs.[8]

Accusations against the Farm Security Administration flew again during a 1943 congressional investigation. Unlike the earlier Byrd committee, the Select Committee to Investigate the Activities of the Farm Security Administration took direct aim at the FSA. The committee's chair was Representative Harold D. Cooley of North Carolina, and Cooley gave FSA opponents free rein to air the familiar litany of charges against the FSA programs and leadership. The Cooley committee also spent a considerable amount of time reviewing the FSA medical care programs, finding widely divergent views of the agency's health programs among physicians. The FSA mustered just enough support to prevent the committee from recommending its total elimination, but this Pyrrhic victory generated little optimism in the beleaguered agency. In fact, Cooley introduced a bill to abolish the FSA and replace it with a new agency, the Farmers Home Administration (FHA), whose mandate would be far more limited and would be more responsive to traditional agricultural interests. Although Cooley's bill was tabled by Congress, he introduced nearly identical bills over the next two years.

Friends of the FSA in Congress, in the Roosevelt administration, and outside of Washington tried to limit the damage being done to it. In 1941, Congressman John Tolan of California chaired a committee that examined farm labor issues as they related to the war effort. The House Select Committee to Investigate National Defense Migration provided a friendly forum for FSA supporters to make their views known. Liberal farm organizations, such as the National Farmers' Union and Southern Tenant Farmers Union, as well as influential Christian leaders, made repeated trips to Capitol Hill to testify on behalf of the agency. Southern New Deal Democrats and politicians from states with large migrant communities—such as senators John Bankhead (Alabama), Lister Hill (Alabama), Richard Russell (Georgia), Sam Rayburn (Texas), and Claude Pepper (Florida)—also stood by the FSA. During the 1942 budget debate, Bankhead and Russell admonished their colleagues for their intemperate and scurrilous attacks on the character and patriotism of Baldwin and other FSA leaders. Support also came from the Roosevelt administration, most notably from first lady Eleanor Roosevelt, who lobbied her husband to do more to protect the agency.[9]

In the spring of 1943, soon after Congress transferred the migrant programs to the War Food Administration, WFA administrator Chester Davis commissioned a confidential and independent assessment of the FSA. Davis, who was not considered a reliable friend of the FSA, asked Harvard economist John D. Black to chair the group. Several months later, the committee submitted a report that acknowledged the substance of some of the criticisms directed at the FSA. Fortunately for the FSA, the report noted that many of these criticisms were exaggerated, and it concluded by calling the FSA "one of the most significant social inventions developed in the field of agriculture in recent decades."[10]

FSA administrators took steps to shore up political support by tailoring the agency's rehabilitation programs to fit better with the country's wartime priorities. The FSA gave defense preparedness classes and collaborated with the Red Cross to teach first aid in federal and private migrant camps. Similarly, the Health Services Branch collaborated with the U.S. Public Health Service and state and local health departments on venereal disease control programs in farm labor camps. This effort was aimed equally at protecting the health of an essential work force and at limiting the opportunity for male army recruits isolated in army camps throughout the South to contract venereal disease. Allen Koplin, a U.S. Public Health Service medical officer, recalled a meeting he had with an army physician in which he hoped to get venereal disease treatment for the farm workers in the camps under his jurisdiction:

> "It's very simple, Colonel. My guys are going to the same whore as your guys are going, and if they pick up the organism and spread it around, it's going to be harder for your guys to keep clean. I want to ask you whether you'll let my fellows go to the 'pro' station." The administrative types said, "You can't do it, military regulations, they're civilians." I said, "Not really, they're helping the war effort too, away from home, like your guys." The colonel said, "When my military judgment comes into conflict with my medical judgment, I always take the latter. Go ahead." And that's the story about how the FSA cooperated with the army. I never told that story to anybody before.[11]

The FSA also put its Information Division to work pitching the value of the nation's small farmers to wartime food production, highlighting the fact that FSA families outproduced those farmers not receiving federal assistance. According to a widely distributed 1944 FSA pamphlet, good health was vital to the nation's capacity to win the war. "If you found a Nazi saboteur on your farm," noted the pamphlet, "you'd get rid of him in a hurry, wouldn't you? There may be other things that are just as dangerous to your health and well-being— and to the war effort."[12] Echoing the FSA's informational campaign, medical journals noted that 85 percent of all FSA borrowers paid off their loans. Similarly approving comments were made when it was learned that Selective Service rejection rates were 30 percent lower in rural communities where the FSA was involved than they were in communities without FSA assistance.[13] In the context of a nation at war, these statistics added support for the FSA in the rural medical community and lent credence to the FSA's motto that rehabilitation made good economic sense.

There were other patches of sunlight, admittedly small and temporary, for the FSA medical care programs during this period of institutional retrenchment. For example, although the transfer of the farm labor programs to the War Food Administration underscored the FSA's declining status in Washington, the reorganization's impact on the migrant health programs was less immediate. This was due largely to the labor shortages caused by the war and by the shortage of health personnel.

Loss of farm labor at a time when efficient harvesting of the nation's crops was essential to the war effort led the federal government to undertake an extraordinary plan of agricultural mobilization. FSA migrant camps were transformed into staging areas from which agricultural workers were assigned to harvest the fields of local commercial growers or were transported from one part of the country to another as seasonal harvest needs dictated. In addition, the United States brought in some 300,000 foreign farm workers from Mexico, Newfoundland, Barbados, the Bahamas, and Jamaica as part of the Emergency Farm Labor Shelter Program.[14] The FSA's involvement since 1937 in transporting, sheltering, feeding, and providing medical care to migrant workers meant it had the necessary experience to assist in this effort, and until its duties were transferred to the War Food Administration, it shared the responsibility of caring for and deploying these workers with other federal agencies.

Treaties between the United States and its neighbors to the north and south specified that imported farm workers would receive medical care in addition to food, shelter, and wages. The duties of the FSA Health Services Branch expanded along with the Emergency Farm Labor Shelter Program. The paucity of available health personnel in government service led to a pragmatic decision by which FSA Health Services staff were given joint appointments in the WFA. Fred Mott, for example, was assigned double duty when he was named as chief medical officer for the War Food Administration in July 1943.[15] As the number of farm labor camps grew, the migrant health programs expanded. By 1945, the farm labor medical programs operated to some degree in virtually every state as well as Puerto Rico.

Whenever feasible, private practitioners provided medical care to both domestic and imported farm laborers as part of the Emergency Farm Labor Shelter Program. As the scarcity of rural health personnel worsened, however, more and more U.S. Public Health Service nurses and physicians were assigned to the farm labor clinics.

The importance of food production to the war effort was not the sole reason the migrant health programs were largely spared during this period of government cuts. Many rural communities, state and county medical societies, and individual physicians urged the government to continue the role it first assumed when localities and state communities had neither the resources nor the desire to accept responsibility for migrants' welfare. The medical and public health programs centered around the federal migrant camps were popular because they lessened or prevented disease outbreaks that had often accompanied the seasonal influx of migrants. Even after the economy regained its footing and replenished the once-barren coffers of local and state governments, rural communities and physicians worried that they would once again be left to shoulder the burden for migrants' health and welfare alone, a concern that was amplified by the importation of farm labor throughout the war. In 1943, Congress passed legislation cutting federal support to the migrant health program and restricted medical care coverage to workers recruited, transported, or placed by the War Food Administration. The law prompted a loud outcry from rural communities and county medical groups in states with large migrant populations. "Current attacks on the FSA threaten to undo one

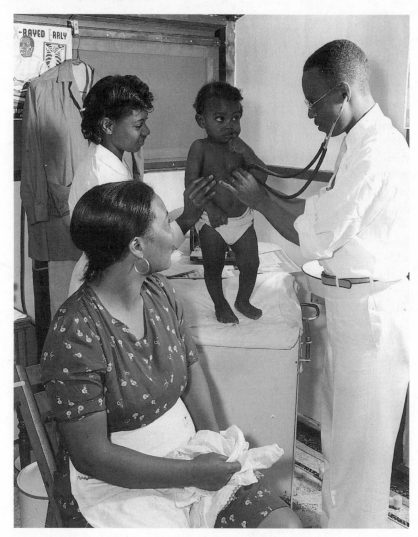

Farm Security Administration agricultural workers' camp. The camp doctor has a large practice in Bridgeton but also gives his time to the migrant camp, where he works to curb venereal disease, malnutrition and general rundown conditions of health. Bridgeton, N.J., June 1942. John Collier.

of the most effective programs of the New Deal," *California and Western Medicine* noted. "The farm bloc, now running wild in Congress, probably will sweep aside this aid to rural communities."[16] Although the bill was backed by the conservative agricultural establishment, New Deal politicians and liberal newspapers joined forces with medical societies and forced Congress to rescind its decision.

CONSIDERING ALTERNATIVES:
ORGANIZED MEDICINE'S VIEWS ON THE FSA

Until 1943, the American Medical Association's public posture toward the FSA was ostensibly neutral. Unwilling to enter into a formal agreement on behalf of the medical profession, the AMA left the decision to its constituent medical societies and individual doctors. Despite its strong stand in favor of fee-for-service solo practice, the organization needed to be attentive to the concerns of rural doctors. But beneath its studious neutrality lurked a deeply felt anxiety about the threat posed by the FSA medical care programs. In 1937, the AMA had warned that the growing federal presence in health care and alliances developing within the federal bureaucracy might establish undesirable precedents from which the profession would never escape. "Regimentation, regulation, red-tape, limitation of choice, compensation by salary, and administration by nonmedical personnel are characteristics of systems of state managed medicine. These characteristics may not all appear at once but each new proposal should be examined with great care to detect tendencies, applications, and hidden phrases which may have grave effects on the future of medicine."[17] A few years later, Morris Fishbein, the influential editor-in-chief of the *Journal of the American Medical Association,* spoke more directly to the FSA program: "We are experimenting. We are trying out a lot of new plans. They seem marvelous before the game starts, but when the game starts you suddenly find blockers appearing out of nowhere . . . Therefore whenever a plan like the FSA plan is set up in one state, that is no reason why 47 other states should immediately adopt the plan."[18] Behind the scenes, the AMA often worked at cross-purposes to the FSA. This was certainly the case in the Dakotas, where AMA machinations delayed the negotiations that ultimately led to the farmers mutual aid corporations.

For the most part, however, county and state medical societies welcomed the FSA programs. By 1942, physicians and medical societies in at least 41 states had witnessed seven years of experimentation in rural health care delivery under the banner of the federal government. Much to the AMA's chagrin, the experience of physicians and county medical societies with the FSA program was largely positive. This was demonstrated most clearly by the results of the AMA's own survey of over 1,000 county and 36 state medical societies in 1941. Participating physicians approved of the FSA program by a decisive 4:1 ratio. An equally large percentage felt that FSA families received better medical care as a result of the medical care program and that physicians made fewer house calls, built their own practices, and earned more money while participating in the FSA plans.[19] Interestingly, many rural physicians valued the program because it gave them firsthand experience with group prepayment insurance and reinforced the perception of the medical profession

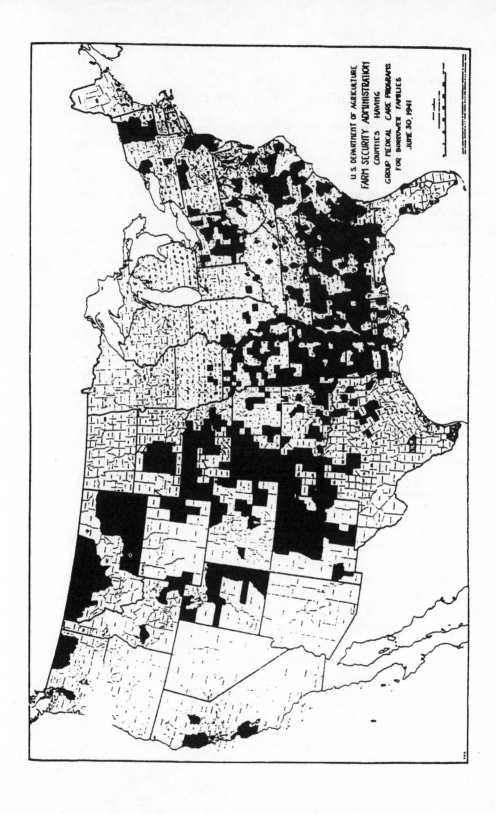

U.S. DEPARTMENT OF AGRICULTURE
FARM SECURITY ADMINISTRATION
COUNTIES HAVING
GROUP MEDICAL CARE PROGRAMS
FOR BORROWER FAMILIES
JUNE 30, 1941

as guardian of the public's health. For these individuals, the FSA plans were a glimpse of the future and it was not as onerous or threatening as the predominantly urban, specialty-oriented, and more prosperous colleagues who spoke for organized medicine feared it would be. "We are convinced that the farm plan is filling in a very definite need," said one New Jersey physician in his testimony before Congress. "We are learning something about costs of rural medical care, developing administrative methods and demonstrating the will of the profession to distribute medical services to all persons at a cost they can afford to pay."[20]

Not all physicians felt this way, of course. As the economic climate improved, rural physicians were less financially dependent on the FSA plans than they had once been. Rising incomes fueled an increased demand for medical services by the public and enabled more and more families to pay their bills. Nationally, annual income for physicians climbed 300 percent between 1939 and 1946. The FSA program was "tempting because it offers you cash," a physician noted in a 1940 *Journal of the American Medical Association* article. But, he added, "it does not offer you any cash you can't do without."[21] In a 1942 article, a South Dakota physician suggested that the return of prosperity to the Dakotas made the statewide FSA-sponsored medical care plan less necessary than had been the case "during the years of drought and grasshoppers."[22] While similar comments were heard during the worst years of the Depression, the FSA's success in recruiting rural physicians to its programs indicates that this was not the dominant opinion.

The national emergency also tempered attitudes toward the FSA medical care plans. Patriotism and the importance of food production to the war effort encouraged some county medical societies and practitioners to be more judicious in their criticism of the agency. The overall atmosphere was one in which physicians expressed greater resistance to continuing their involvement with the FSA, but this was not uniformly the case. In some rural communities, doctors were so busy that they were more willing than before to let the government provide support for migrants and other low-income families. A number of county medical societies renegotiated lapsed memoranda of understanding with the FSA following the declaration of war. The FSA also appealed directly to physicians' sense of duty, and emphasized the small farmer issue to elicit physicians' support. Judging by the tone of comments on the FSA in medical journals during the war, this strategy was moderately successful.[23] The Medical Association of the State of Alabama reported that "the FSA families have responded 100% to the call . . . We trust that through the cooperation of state and county medical societies and participating physicians that [*sic*] we

Facing page: *Counties with medical care programs in place as of June 1941.*
Annual Report of the Chief Medical Officer, *1941.*

Doctor Onstine, MD, writing a prescription in the clinic operated by the Taos County Cooperative Health Association. Questa, N.M., January 1943. John Collier.

may have that cooperation during the time to come when health conditions are going to play a large part in producing food for our fighting forces."[24]

The AMA's reluctance to be oppositional at a time when healthy hands were needed for healthy harvests was evident in medical journals at the time as well. The AMA wanted to avoid acting in ways that might be construed as unpatriotic and self-serving, particularly after the organization politically blundered by opposing the Emergency Maternal and Infant Care (EMIC) Act when it was first proposed in 1942. The EMIC program required doctors to provide obstetrical, postnatal, and early childhood services to the families of infantrymen in the first four pay grades. Initial sentiment among many doctors and organized medical groups toward the bill was strongly negative.[25] When it was clear that the bill was wildly popular with Congress and the public, organized medical groups first muted and then withdrew their opposition. Chastened by this experience, the AMA was careful not to allow organized medicine to appear as if it was unwilling to shoulder its share of the home front's war burden.[26]

The FSA's new-found approbation was also visible in statements made at the AMA house of delegates meetings in 1941 and 1942. At the 1941 meeting, delegates were told that the FSA was "becoming broader and more effective each year" and that the "rehabilitation work done under the supervision of Dr. Williams has been a valuable contribution."[27] This apparent change of heart gratified the FSA Health Services staff. FSA leaders believed all along that their programs improved the quality of medical care in rural communities and that they bolstered rather than undermined the status of the medical profession. In 1941, the agency's chief medical officer wrote: "There is increasing approval of the program on the part of organized medicine, an acknowledgment that the organization of payment for medical service need not interfere with the physician-patient relationship."[28]

There were other considerations that led organized medicine to be more circumspect in its public posture toward the FSA. Many physicians and medical societies reconsidered their attitude toward the FSA against a backdrop that included the recommendations of the Interdepartmental Committee to Coordinate Health and Welfare Activities, the 1938 National Health Conference, and various national health bills introduced in Congress from 1939 to 1946. There was also doctors' nagging suspicion that Roosevelt still hoped to add national health insurance to the New Deal's legislative legacy. When viewed in the context of these developments, and as compulsory insurance schemes gained wider currency in the mid-1940s, it was not surprising that for some physicians the FSA medical care programs assumed a more modest appearance. As A. M. Simon of the AMA acknowledged in a 1942 editorial: "There have been theoretical objections and charges that such plans are a step toward 'State Medicine,' 'compulsory sickness insurance,' or 'socialized medicine' . . . In evaluating such objections, it may be well to consider alternatives."[29]

Closer inspection of medical commentary reveals that this support was to an important degree contingent on the FSA's focus on low-income farmers, the financial benefits gained by rural doctors, and on the FSA's incorporation of key tenets of organized medicine, such as voluntary participation and free choice of physician. There were limits to organized medicine's tolerance, however. When the FSA tested the boundaries of the medical care program—such as loosening eligibility criteria or allowing non-FSA families to obtain care at migrant camp clinics—it provoked greater obstinacy and calls for ongoing vigilance to ensure that the FSA did not stray too far from its focus on the rural poor. At the 1940 AMA annual meeting where delegates were told that the

FSA's "pool plan is practical for physicians and patients," the Committee on Legislation also emphasized that the FSA program "is a specific plan for an underprivileged group of farm people and it would have to have considerable modification and changes if transferred to any other segment of the population."[30] The following year, the AMA house of delegates warned the FSA that any "amendment of the by-laws to include other than FSA clients was contrary to the principles of the AMA and the expressed policies of the FSA to cooperate fully with state and county societies."[31]

August 1943 marked a crucial transition period for the FSA. It began when Beanie Baldwin resigned his post as FSA administrator. Baldwin was genuinely eager to explore new career opportunities in the relief and rehabilitation programs established by the Allies in liberated zones in Europe, but it was not happenstance that he departed the FSA shortly after Congress curtailed the agency's role in the farm labor shelter program. Congress's act confirmed Baldwin's doubts that the FSA's efforts to adjust its rehabilitation efforts to the wartime priorities could stave off further damage to the agency. Like his two predecessors, Baldwin had wearied of the virulent attacks made against the FSA and himself, and he knew that it was only a matter of time before he too would be sacrificed in order to placate the agency's opponents.[32] Baldwin's decision was one of the reasons Congress decided not to take up Cooley's bill to eliminate the agency.

Baldwin was temporarily succeeded by Robert W. "Pete" Hudgens, who functioned as acting administrator for several months. In early November, Hudgens was replaced by Franklin W. Hancock, a moderate southern Democrat serving in the House of Representatives. Hudgens remained on the Washington staff of the FSA until 1946, but few of the agency's old guard stayed on as long. By December 1943, virtually the entire senior FSA leadership in Washington and regional offices had moved on to new positions.[33]

The exodus of senior leadership disrupted the FSA's rural rehabilitation efforts, the medical care programs included. The leadership vacuum created when Baldwin resigned added to the difficulties already facing the medical care program. The agency's chief medical officer considered Baldwin's resignation a potentially "crippling blow" for the FSA, and one that left the future of the medical care program "unsettled."[34]

THE UNHOLY ALLIANCE: THE AMA, THE FARM BLOC, AND VOLUNTARY HEALTH INSURANCE

Broader changes occurring in medical care financing over this same period undermined still further the FSA's ability to defend its medical care programs. Perhaps the most important was the emergence and expansion of private voluntary health plans in rural communities, some of them sponsored by the

FSA's fiercest critics, the Grange and the Farm Bureau. According to a 1938 Gallup poll, nearly 50 percent of all Americans, and over two-thirds of low-income families, had delayed seeking medical care in the last year for financial reasons.[35] The realization that almost half the population, and not just the indigent, experienced economic obstacles to medical services was a persuasive argument for reform. Organized medicine was finally motivated to respond credibly to the fact that for too many Americans medical care was unaffordable and inaccessible.

Social reformers and organized medicine offered radically different solutions to the central issue of how to eliminate financial barriers to a service that the majority of citizens considered to be a public good. The question was not whether change was needed, nor whether access to medical care was a good thing. Everyone with a vested interest in health care assumed that scientific progress had brought real gains to medicine and that access to better medical care was an unmitigated public good.[36]

On one side of the issue were reformers eager to complete the New Deal's social welfare legacy by passing national health insurance legislation. They pushed for financing reform, which they believed would be best achieved through a publicly financed health system. The more militant voices argued for fundamental change in the organization of health services along with universal coverage and public financing. President Roosevelt was unwilling to press the issue, but there were those in government service who did so. Chief among government activists was I. S. Falk of the Social Security Board. A sociologist by training, Falk had been a key staff member of the Committee on the Costs of Medical Care, the Committee on Economic Security, and the Interdepartmental Committee for the Coordination of Health and Welfare Activities. He is widely seen as one of the driving intellects behind the national health insurance movement at midcentury and was the acknowledged author of the various Wagner-Murray-Dingell bills. Also prominent were Surgeon General Thomas Parran, Frances Perkins from the Department of Labor, and Martha May Eliot of the Children's Bureau. Others advocated national health insurance from their positions in private philanthropies; for example, Michael M. Davis of the Julius Rosenwald Fund and Edgar Sydenstricker, who had left the U.S. Public Health Service for the Milbank Memorial Fund. A number of vocal academics, including Nathan Sinai at the University of Michigan School of Public Health and Henry Sigerist at Johns Hopkins, also supported the reform cause. Innovative models sponsored by private philanthropies, industry, and consumer groups emerged, including plans based on cooperative principles, capitation, group practice, and salaried physician staff. And although private fee-for-service practice remained the dominant mode of medical practice, prepaid group insurance and group medical practice were no longer rarities in midcentury American medicine.

Arrayed against the reform camp were organized medicine and the majority of physicians. They were flanked by most groups with a financial interest in health care, including hospitals, the business community, and the insurance industry. Led by the AMA, these groups sought to deflect more radical reform by promoting private and professionally sponsored initiatives, such as Blue Cross and physicians' service bureaus. Private, voluntary insurance plans were generally less inventive in terms of the organization of medical services. By the time the first Wagner-Murray-Dingell national health insurance bill was proposed in 1943, voluntary health insurance had already emerged as organized medicine's buffer against national health insurance. Although the development of third-party health insurance in the United States is well covered in standard histories of American medicine, some details are necessary in order to understand its impact on the FSA medical programs.[37]

The American Medical Association was initially hostile to any outside control of medical care, but above all to government control. When it became clear that some form of health insurance was going to be a feature of the future, the AMA tacked to the left and touted the advantages of private, voluntary plans. The AMA's reversal on third-party insurance in 1938 freed physicians and medical societies to take greater advantage of the public's interest in group prepayment insurance. By supporting those initiatives that promised to be most conservative and least intrusive, the AMA correctly anticipated that expanding private health insurance would undermine popular and political support for national health insurance. "We all hope," wrote an Arkansas physician, "that voluntary payment plans, sponsored either by medical societies or commercial insurance—will meet the palpable demand of the public for relief from the unpredictable financial burdens of illness. Once firmly established, they would set the pattern."[38]

It was a winning strategy. The medical profession's willingness to consider ways to meet the problem of cost and access barriers to medical services and the public's response to these initiatives drove the growth in third-party health insurance throughout this period. This in turn deflated the arguments of those who supported more fundamental changes in medical care and further consolidated the profession's authority over the practice of medicine.

From 1933 to 1942, the percentage of Americans with some form of health insurance more than tripled, going from 6 percent to 20 percent.[39] Most were covered by commercial indemnity insurers, group hospital insurance plans such as Blue Cross, and physicians' service bureau plans promoted by state medical societies. In general, these group prepayment plans emphasized coverage for the costs of hospitalization over the payment of office care. Although a third party was now present in the doctor-patient relationship, voluntary health insurance reinforced the dominance of fee-for-service allopathic medicine under the control of organized medicine. These plans were most often available as a benefit of employment.

Rural communities' interest in group prepayment insurance also blossomed during this period, due in no small part to the emphasis given to the issue by the FSA. The FSA's successful development of a prepaid group insurance program, and the attention the agency brought to rural health issues in general, spurred organized medicine to address the health concerns of rural communities. More than thirty state medical societies and the AMA had standing committees on rural health by the beginning of World War II. There were those in the medical community who credited the FSA for bringing the issue of rural health and medical care to the forefront of national debate. In 1946, for example, the California Medical Association credited the FSA for a "daring venture" that had caught the eye of farm groups and medical associations nationwide.[40]

Without question, the FSA's involvement in rural health care encouraged agricultural organizations to address rural health issues. Of the three major farm groups, only the National Farmers' Union demonstrated any serious commitment to rural health issues prior to the New Deal era. The National Farmers' Union was one of the FSA's most consistent allies in the agricultural establishment. Not only did the liberal farm group support the FSA programs, it endorsed a variety of health care reform proposals during the 1930s and 1940s, national health insurance included.[41]

The Grange and American Farm Bureau Federation were slower to address the issue of rural health. Some observers at the time, including many in the FSA Health Services Branch, suspected that the interest of those organizations in health insurance was motivated as much by resentment of the FSA as by a genuine desire to meet the needs of their own members. Helen Johnson, who worked in the Department of Agriculture's Rural Cooperative division and who knew many of the FSA staff, believes that the Grange and AFBF proposals were "sort of a strategy by which this could be kept in more private hands."[42] By the early 1940s, however, both groups were promoting voluntary health insurance plans for their members, often with the enthusiastic support of organized medicine.

The eagerness with which organized medicine embraced the health insurance proposals of conservative farm organizations was not simply due to changing professional attitudes on group health insurance. Something more fundamental was at work. There was a natural chemistry between organized medicine and commercial farmers that was based on common values. The leadership of the AMA was made up almost exclusively of more well-to-do urban specialists, while the Grange and the Farm Bureau represented more affluent commercial farmers. The AMA portrayed itself as the defender of the solo fee-for-service physician, while conservative farm groups protected the rights of the self-reliant farmer. Organized medicine and organized agriculture exalted individualism and freedom over collectivism and regulation, and for that reason alone they would view the FSA's efforts as largely inimical to the interests of their membership. This shared sense of identity was perhaps best expressed by the *Journal of the American Medical Association* when it observed that farmers and physicians were by their very nature "self-confident individualists" who were "impatient with restraint."[43]

The bond between organized medicine and the farm bloc offered political reciprocity as well. Organized medicine and conservative agricultural groups distrusted the thrust of the New Deal, and each group viewed national health insurance as an unwarranted and dangerous expansion of government authority. This point was driven home in a 1945 *Journal of the American Medical Association* article: "With the one exception of the Farmers' Union, all farm groups have adopted a constructive attitude toward . . . compulsory socialized medicine. We as a medical profession have many links in common with them, many areas of agreement. They are our natural allies in the fight to preserve this nation from regimentation, and it is to our mutual advantage that we work in close harmony." [44]

In the farm bloc, organized medicine found a willing confederate in the fight against socialized medicine. In the medical profession, the farm bloc had a useful ally in their efforts to cripple an agency that challenged its traditional authority in rural communities.

Agricultural organizations emphasized group hospital insurance, rather than payment for physicians' services provided in the office or in the home. In this respect, the growth of rural health insurance followed the pattern of the nation as a whole. Typically, Farm Bureau or Grange members enrolled in existing Blue Cross hospital insurance plans, or in prepaid group insurance plans sponsored independently by the AFBF or the Grange. Medical services outside of the hospital were much less commonly available, but cooperative arrangements with physicians' service bureaus or separate Grange and Farm Bureau indemnity plans were sometimes put into place.[45]

Although the insurance options available to rural families grew throughout the 1940s, the growth of prepaid group insurance in rural America still lagged behind that of nonrural communities. Lingering problems such as poorly trained medical personnel, lack of accessible physicians, inadequate health infrastructure, poor sanitation and nutrition, and low baseline health status remained major barriers to improving the health of rural citizens. These chronic problems were quite familiar to those who worked for the Farm Security Administration. In addition, wide variations in enrollment and significant membership turnover in the FSA medical care plans also suggest that not all low-income rural families were convinced of the benefits of health insurance.

By 1941, less than 3 percent of farmers had any kind of health insurance, whereas 10.5 percent of the nonfarm population and nearly 20 percent of urban families were insured to some degree. Similar discrepancies between urban and rural areas were seen for hospital insurance. Of the 24 million Blue Cross hospital plan members in 1947, for example, only 1.6 million were classified as rural. These figures were only marginally higher well into the next decade. Commercial indemnity plans and physicians' service bureaus fared little better. In 1946, only a small fraction of the 8 million individuals enrolled in plans offered by private insurers were from rural counties. In 1947, physicians' service bureaus boasted a membership of some 6 million, but most members were drawn from industrial communities in Michigan, California, Oregon, and Washington. Rural membership in physicians' service bureaus stood at less than 7 percent.[46]

The growth of private and physician-sponsored voluntary health insurance plans was problematic for the FSA's medical leaders. Organized medicine's enthusiasm for their own medical service plans, or for the plans developed by conservative farm organizations, did not always immediately disrupt established FSA plans. Still, once state medical societies began their own group prepayment plans, they often pressured the FSA to enroll its rehabilitation borrowers in them. In more than one state, the infusion of FSA clients into fledgling physicians' service bureaus gave a much-appreciated boost to the medical society's plans.[47] In California, a trial enrollment of several hundred FSA rehabilitation families in the California Physicians Service began in 1941. The pilot program was so successful that plans were made to expand it statewide the next year, making it possible to enroll nearly 35,000 FSA rehabilitation families in this physicians' service bureau. Similar trends were seen with hospital coverage for FSA families. Between 1941 and 1945, the FSA paid to enroll its clients in group hospital insurance plans (including several Blue Cross plans) in at least seven states.[48]

FSA medical leaders observed these developments with a mixture of resignation and apprehension. The leveraging of FSA clients into physicians' service bureaus and hospital insurance plans, which emphasized organized medicine's growing strength and the FSA's declining influence, led some FSA personnel to conclude that private insurance plans might be the only option left for low-income farmers. At a national meeting of hospital administrators in 1944, the chief of the agency's Health Services Specialists bureau, Ken Pohlman, admitted that although the FSA program was a "great benefit" to FSA families, it remained "inadequate, incomplete, and fragmentary."[49] Citing the growing antipathy to the FSA, Pohlman ruefully voiced his hope that physician-sponsored plans might fill the gap being created as the agency slowly hemorrhaged.

Most FSA Health Services staff had more serious reservations about these developments. For one thing, the plans being touted by organized medicine and their agricultural allies were rarely as comprehensive as the FSA's own plans. These alternative plans offered hospital coverage, but few covered ordinary physicians' visits and none incorporated the broad range of nursing, prevention, health education, nutrition, and sanitation services typically available to FSA families. Even in physicians' service bureaus with significant numbers of FSA members, the agency's efforts to encourage the sponsoring medical society to include anything other than medical services met with little success.[50] There was also the very real concern that private insurance plans were too costly for most low-income families to join, at least not without government subsidies. More to the point, these plans were not focused primarily on attracting poorer families as members. Finally, insurance plans controlled by or closely tied to organized medicine evinced little interest in altering either the organization or the practice of medicine, two issues that the FSA had demonstrated its willingness to tackle.

Subsequent experience validated many of the concerns raised by the FSA's medical staff. Private group health insurance plans made few efforts to enroll low-income rural families. What gains private and professionally sponsored health insurance plans made in rural areas during this period were limited in general to higher-earning rural families and emphasized hospital coverage. In 1946, less than 4 percent of low-income families in the twenty-eight most rural states had Blue Cross coverage. In states where there were mergers between the FSA and state medical society plans, the agency had to subsidize annual membership dues for its clients. The AMA itself estimated that less than 15 percent of rural families could afford to join private insurance programs without direct government support.[51]

Organized medical groups and physicians were not very concerned about the issues raised by the FSA's medical leaders. One reason was the profession's growing confidence in the public's preference for voluntary health insurance, or at least in doctors' ability to convince them that it was best. Even in those states where the FSA enjoyed a solid working relationship with organized medical groups, most doctors believed that FSA families would abandon the government plans in favor of physicians' service bureaus or Blue Cross plans if given the choice. With few exceptions, the agricultural establishment wholeheartedly supported organized medicine's contention that the growth of private and professionally sponsored health insurance plans made the FSA's involvement in rural health care delivery unnecessary. Together, organized medicine and organized agriculture emphasized that rural citizens were able to join voluntary group prepayment plans, and that a growing number of them were opting to do so.[52] The fact that low-income families were the least capable of taking advantage of the growing number of insurance plans was of less consequence.

This sentiment was not limited to anti–New Deal politicians, conservative farm organizations, and organized medicine. The Interbureau Committee to Coordinate Post-War Programs had itself discouraged significant expansion of federally sponsored medical care programs in rural communities, despite the reservations of FSA representatives, because a majority on the committee believed that private health insurance made federal programs expendable. Les Falk (no relation to I. S. Falk) later acknowledged the adverse impact private insurance plans had on the FSA programs. "There was an increasing perception that the private sector—in the form of Blue Cross, the American Farm Bureau or Grange supported prepayment plans, or all those physicians' service bureaus that were starting to crop up—would fill the gap that the Farm Security plans had started and that it was unnecessary to have a federal counterpart to that."[53]

The emergence of private health insurance, along with the easing of the Depression and the coming of World War II, created a situation in which the FSA was no longer able to hold its own against the forces of money and power arrayed against it. Senior FSA Health Services Branch personnel viewed the relationship between organized medicine and conservative farm groups as an especially worrisome development. FSA Chief Medical Officer Fred Mott considered this "unholy alliance" a dire threat to the agency's medical care programs.[54] This was the context in which the FSA administrators made the decision to embrace publicly the cause of national health insurance. Even had FSA leaders been inclined to stand on the sidelines of the controversy, organized medicine's view of the medical care programs and the close ties that agency leaders had to high-profile government reformers would surely have drawn them into the debate.

Although the FSA was already in decline by the time the agency's leaders spoke out in favor of national health insurance, this fact did not lessen organized medicine's consternation over the agency's decision. AMA leaders had warned of the threat posed by the FSA before, but now the agency's visibility in the national health movement confirmed the AMA's suspicions about the agency's true motives. From this point on, organized medicine connected the FSA medical care programs to the national health insurance movement, and portrayed members of the Health Services Branch as part of a cabal of government physicians and social reformers conspiring to impress socialized medicine on the American people and the medical profession. As organized medicine stepped up its campaign against the FSA medical care programs, it was wholeheartedly supported by conservative farm interests. The FSA's decision to champion national health insurance also drove rural physicians into the open arms of those who had been clamoring for the termination of the agency for years, subverting physicians' support at the local level just as the AMA declared that the apocalypse—in the form of compulsory national health insurance—was upon the profession.

CHAPTER SIX

Running against the Tide
The FSA and National Health Insurance

*The drive for better rural health is right at the forefront of the whole
movement for a national health program . . . We can have the satisfaction
of spearheading a movement which will ultimately benefit every citizen.*
FREDERICK DODGE MOTT, SASKATCHEWAN, CANADA, 1946

The history of national health insurance in the United States is a story of
promise and dashed expectations, power politics and political ineptitude. It
has proven an irresistible attraction to scholars.[1] Establishing a comprehensive social insurance system that included medical care has been among the
most cherished goals of social reformers in the United States since the turn
of the century. While some historians have held that reformers' belief that national health insurance could be instituted was an illusion, others have argued
that victory at various times had been close at hand when it was snatched away
by the forces of reaction in the medical profession. The best opportunity to
enact national health insurance probably occurred in the period that began
with the first Roosevelt administration and ended conclusively in 1946 when
Congress reverted to Republican control. By the end of that decade, the national health insurance movement had suffered an irretrievable loss of momentum. With the ascendance of employment-based insurance, and a private
insurance industry firmly tied to the medical profession, the possibility of a
publicly financed health care system faded, perhaps for good.[2]

FSA medical leaders were unable to avoid being drawn into the maelstrom
of national health politics, and the agency's involvement in medical care delivery was taken seriously by the central actors on both sides of the debate.
While advocates of national health insurance considered the agency's experience valuable to Congress as it wrestled with health care legislation, organized
medicine increasingly portrayed the FSA programs as a dangerous precedent
and a potential avenue through which a federal system of health care might
advance.

The decision of the FSA staff to abandon their previous reticence and speak
out in favor of national health insurance was costly to the agency. The oppo-

sition of rural physicians and county medical societies to national health insurance was no less firm than the AMA's and as the FSA's position on the issue became known, they severed their relationship with the agency. There were personal and professional repercussions as well. The FSA's association with the "failed crusade" damaged the professional reputations of several high-ranking Health Services Branch physicians and drove a number of them out of government service altogether. Why and how the fate of the FSA medical care programs became intertwined with that of the national health insurance movement is the subject of this chapter.

SPEARHEADING A MOVEMENT? THE FSA AND THE 1943 WAGNER-MURRAY-DINGELL BILL

National health insurance was considered, and rejected, by several prominent health policy groups in the 1930s and 1940s, including the Committee on the Costs of Medical Care, the Committee on Economic Security, and the Interdepartmental Committee to Coordinate Health and Welfare Activities. The 1938 recommendations of the Interdepartmental Committee's Technical Committee on Medical Care, however, brought the issue to the forefront of political debate, where it stayed for the next decade. From 1939 to 1949, Congress considered a range of bills addressing health care. Some of these proposals were quite modest; others were anything but.

One of the earliest of these bills was the 1939 National Health Bill introduced by Senator Robert Wagner. The Wagner bill called for federal dollars to be provided to state governments for the creation of medical care programs for the general population. The proposal left the details of how these programs were to be run to the individual states. The bill also proposed amending the Social Security Act to provide medical care for mothers and infants, enlarging the programs of the U.S. Children's Bureau, increasing funding for the U.S. Public Health Service and for state public health programs, providing money for hospital construction, and creating federal-state disability coverage.

After the 1939 Wagner bill died in Congress, the next big push for national health legislation came in 1943 when the first Wagner-Murray-Dingell bill was introduced. This bill proposed adding universal health insurance to the existing old-age, unemployment, and disability insurance programs created by Social Security. Financing for the health insurance fund would come from increased payroll deductions, and the fund was to be administered centrally by the federal government. Administratively and in terms of funding, the 1943 Wagner-Murray-Dingell proposal represented essentially an extension of the popular Social Security system.

The public circumspection of FSA leaders on national health insurance ended with the introduction of the bill. Up to this point, the agency's official position was that its medical care delivery program was grounded in economic rather than in ideological or political goals. However, the FSA's medical leaders had always been interested in more than simple loan protection. "Some years ago there was a tendency for us to avoid publicity because of difficult professional relationships," wrote Chief Medical Officer Fred Mott in a confidential 1943 staff memorandum, "but I feel strongly that time has passed."[3] When Congress summoned Mott to present the FSA's position on the Wagner-Murray-Dingell proposal, he wrote to his colleagues that "the request raises an important question about the dilemma—which is not new to any of us—that of working for certain goals privately and yet having to take carefully considered stands publicly."[4] Despite his reservations, however, Mott favored the legislation and said so in public testimony before Congress.

In later years, participants in the FSA medical care program admitted that they had soft-pedaled their interest in national health insurance. "It was certainly downplayed," recalled Les Falk, "because the idea was to develop more of a private-public partnership to help the program expand. And it was a fairly successful strategy, but in the end it didn't get them anywhere. Then they became more publicly aligned with the Surgeon General, and then obviously this allowed the programs to be further attacked."[5]

FSA administrators brought the agency's considerable expertise in education and public relations to bear on behalf of the Wagner-Murray-Dingell bill. For example, the agency's Information Division produced brochures describing the bill in highly favorable terms. FSA leaders were well aware that this promotional effort put them in the awkward position of arguing that the health needs of rural America were significant enough to warrant a major new federal program without implying that the agency's own efforts were ineffective.[6]

The timing of the FSA's decision requires explanation. The agency had just (barely) survived a series of frontal assaults on its budget and programs when the Wagner-Murray-Dingell bill came before Congress. In light of its political troubles, why would the FSA risk the support of the medical community that it had worked so diligently to accommodate? Did FSA medical leaders misjudge the effect of their decision on physicians, or did they overestimate the support that national health insurance had among rank-and-file physicians and the public? Or, given the internal difficulties with physician recruitment, membership turnover, and the emergence of private or physician-sponsored alternatives, did FSA medical leaders decide that they had little to lose?

FSA leaders were motivated by both practical considerations and ideology. The flaws of voluntary strategies as a means of extending health insurance coverage to rural communities were apparent to the FSA's senior medical leaders by the early 1940s. Without compulsory enrollment, higher-income families would opt out of group prepayment plans, placing such plans on shaky actuarial ground.

The growth of private and physician-sponsored group insurance plans—including those sponsored by the FSA's nemesis, the farm bloc—added to the quandary facing the FSA Health Services staff. They knew that commercial insurers, physicians' service bureaus, and conservative farm organizations were not interested in the most financially vulnerable rural families. Even had insurance programs sponsored by these groups been eager to attract low-income families, few such families could afford to join them without government subsidies. The FSA did enroll some of its clients in alternative insurance plans in a number of states. However, enrolling most of its clients in private insurance plans would have required a level of congressional support that had long since evaporated. Given the limitations of voluntary approaches and diminishing congressional support for social welfare programs in general, compulsory health insurance options became all the more appealing.[7]

Although the FSA's political troubles had dampened the optimism of the agency's staff by this time, the agency's decision to support national health legislation was at least in part driven by a streak of idealism. Battered but not yet beaten, the staff were not always inclined to let prudent politics stand in the way of a good battle for what they thought was right. As one Washington staff member recalled, "There was about us a touch of that famous Roman salute, 'Hail Caesar, we who are about to die salute you!'"[8] In spite of the agency's political troubles, its leaders still believed in medicine's ability to contribute to a better and more humane society and in constructive government intervention to make it available to all citizens. As one of Fred Mott's FSA colleagues recalled: "Mott believed that the FSA had some real significance for the future of medical care. I think that he, and many of the people he worked with, anticipated that national health insurance was not too far away."[9] Forty years later, a bemused Lorin Kerr remembered vividly the optimism that fueled the exertions of medical reformers of the era: "I became convinced that the payment of doctor and hospital bills was the prime health problem confronting the nation at that time. And there were some folks that were like-minded, why maybe we could solve it in a few years!"[10]

Mott and his colleagues were not alone in viewing the agency's medical care programs as a "dry run" for national health insurance. In 1938, the popular magazine, the *Saturday Evening Post,* characterized the FSA programs this way:

> Working with unaccustomed modesty and publicity shyness, the FSA has in effect staged a gigantic rehearsal for health insurance. It has brought together some 3,000 doctors and more than 100,000 families in twenty-odd states. It has given them a chance to show what would happen if a health insurance law were enacted for them tomorrow. And the performance has been truly startling. Friends and foes of socialized medicine alike will be surprised. Our newly elected seventy-sixth Congress may be asked to decide whether this country wants some form of state medicine, but the farmers have it.[11]

Public opinion polls appeared to substantiate reformers' confidence about the imminence and the inevitability of national health insurance. A widely quoted 1944 survey by the National Opinion Research Center indicated that more than two-thirds of Americans agreed that Social Security should be modified to provide universal medical and hospital insurance. Support for national health insurance was even stronger in rural communities. A 1943 Bureau of Agricultural Economics survey indicated that 75 percent of rural Americans favored the extension of Social Security to cover health insurance.[12]

Senior FSA Health Services physicians interpreted their participation in health policy discussions at the national level during World War II as a hopeful sign. Although the final proposals of the Interbureau Committee to Coordinate Post-War Programs fell short of the FSA's expectations, the agency exerted a strong influence on the Interbureau Committee and this reinforced their conviction that federal health policy in the postwar period would draw on the agency's experience—at least as it related to rural communities. This persistent optimism is evident in a September 1943 memorandum that Fred Mott sent to his staff. "It is essential that we look for every opportunity this year to consolidate the health programs, and to bring the realization to other agricultural groups and agencies that we are in a position to assist them . . . in working toward the development of a national health program."[13]

Not everyone working for the FSA was sanguine about the prospects for national health insurance or about the relevance of the FSA's medical care programs to postwar national health planning, however. John Newdorp, a district medical officer who coordinated the farm labor medical programs in the South as the war drew to a close, later recalled: "I can remember having a long argument on a trip with Milton Roemer in which he anticipated that within a year after the war was over we would have national health insurance. I took the other side that it would be a long, long time, if ever."[14]

The FSA's endorsement of the Wagner-Murray-Dingell bill eroded much of the good will the agency had built up with rural physicians and county medical groups. Participating physicians and medical societies felt betrayed, and to the delight of the AMA, accused the FSA of disguising its real agenda all along. In 1943, the state medical society in Texas singled out the Wagner-Murray-Dingell bill, the U.S. Public Health Service, the FSA, and the EMIC program for specific criticism. The organization claimed that the federal government was making "an obvious approach to socialized medicine in Texas," and delivered a strongly worded advisory to county medical groups urging them "*not* to sign or agree to any plan, agreement, or contract sought for the purpose of providing hospital and medical care for any designated group or groups."[15] Similar charges were hurled at the FSA in North Dakota and South Dakota, and complaints against the FSA's political activities featured prominently in a number of state medical journals around the country.[16]

The 1943 Wagner-Murray-Dingell bill was never voted out of committee. The AMA rallied its constituent state and county medical societies in an intense lobbying effort against it. Organized medicine had the support of the commercial insurance industry, hospital associations, conservative farm groups, and much of the business community, notably the powerful Chamber of Commerce. Together, these groups formed a potent, well-financed, and unified alliance.

There were other factors that added to congressional inertia over the issue of national health insurance. In the 1942 midterm elections, voters returned to Washington a Congress that was more determined than ever to rein in the excesses of the New Deal. Roosevelt's flagging commitment to New Deal social programs also contributed. "Dr. New Deal" had given way to "Dr. Win the War" and the president was once again unable or unwilling to risk political capital on behalf of the controversial legislation. "We just can't go up against the State Medical Societies," Roosevelt was quoted as saying. "We just can't do it."[17] While the president's irresolution dismayed the bill's supporters, even congressional liberals were hesitant to press the issue at a time when the war demanded political harmony and domestic belt-tightening.

The reform camp split over the details of the legislation as well. Most supporters of national health insurance agreed that public financing of health care, which the bill incorporated, was a necessary step. However, several prominent reformers, among them Michael M. Davis, expressed doubts about the proposal because it made no attempt to restructure the medical care delivery system. In general, the more militant voices in the reform camp considered the 1943 Wagner-Murray-Dingell too incremental and overly solicitous of organized medicine's interests.

Bickering between federal agencies and key administration figures added confusion to the debate. Prominent government physicians whom reformers had presumed would back the bill, such as Martha May Eliot and Thomas Parran, vacillated when asked to present their views on the Wagner-Murray-Dingell to Congress. Eliot, who was assistant chief of the U.S. Children's Bureau and director of maternal and child health programs, supported those provisions that would have expanded these programs. However, she was critical of other features of the bill. Parran supported the concept of universal health coverage, but he was less enthusiastic about the Wagner-Murray-Dingell bill itself. Parran and other influential Public Health Service personnel felt that their agency should have the primary consultative role with the states in the areas dealt with by the legislation. Although Parran was pilloried by the AMA as a proponent of socialized medicine, his tepid response to the 1943 bill contributed to its demise.[18]

A SECOND PUSH:
THE 1945 WAGNER-MURRAY-DINGELL BILL

On April 12, 1945, just a few months after taking his fourth oath of office, Roosevelt suffered a fatal cerebral hemorrhage in Warm Springs, Georgia. To the surprise of many, the new president, Harry Truman, became a vocal supporter of national health insurance. With Truman's encouragement, a second Wagner-Murray-Dingell bill was introduced in Congress. The 1945 version, sometimes called the Truman plan, would have created a compulsory federal health insurance program. The bill proposed paying for the medical expenses of unemployed citizens through general tax revenues, while increased Social Security payroll deductions would cover the expenses for everyone else.

The 1945 Wagner-Murray-Dingell bill was primarily a financing mechanism and did little to modify the existing organization of medical care. While this issue irked some reformers, organized medicine was not one to quibble over such distinctions. The AMA denounced the proposal as frank socialism which, if enacted, would interfere with patients' right to choose their doctors, disrupt the doctor-patient relationship, and enslave physicians.

Farm Security Administration medical leaders were fully engaged in the effort to pass the 1945 Wagner-Murray-Dingell bill. They recalled the agency's regional and district Health Services field staff to Washington in order to develop a strategic plan to organize support for the bill in rural communities.[19] At the same time, the FSA lobbied Congress to take a closer look at the agency's programs as it considered postwar health policy and planning. In testimony before a Senate committee on the 1945 Wagner-Murray-Dingell bill, Chief Medical Officer Fred Mott told the group that "we should be utterly derelict in our responsibilities today were we to ignore the lessons of this large body of experience in health insurance."[20]

There were more than a few who agreed with Mott that the FSA programs were worth careful scrutiny as the Congress grappled with health care legislation. One congressional subcommittee, the Senate Subcommittee on Wartime Health and Education, undertook a detailed analysis of the FSA experience. In 1944, Harold Murray, chair of the Senate Committee on Labor and Education and one of the co-sponsors of the Wagner-Murray-Dingell bills, assigned this committee the task of reviewing the distribution and utilization of the nation's health personnel, facilities, and services. Murray picked an unrepentant New Dealer, Senator Claude Pepper, as chair of the subcommittee. Joining Pepper on the committee were some of the Senate's most liberal voices, including Robert LaFollette of Wisconsin and Lister Hill of Alabama. (Hill would later sponsor perhaps the most influential piece of postwar health legislation, the 1946 Hill-Burton Hospital Survey and Construction Act.) Over

Dr. Rulfa in his office. San Augustine, Tex., April 1943. John Vachon.

a two-year period, the Pepper committee tilled what was by now familiar soil. It revisited the same issues and made many of the same recommendations as had earlier groups from the Committee on the Costs of Medical Care to the Interbureau Committee to Coordinate Post-War Programs.

The Pepper committee's interest in the FSA medical care programs, in particular the agency's experimental health plans, rekindled expectations within the agency that its efforts would influence postwar federal health policy. With the help of staff from the FSA Health Services Branch and the Bureau of Agricultural Economics, the Pepper committee produced a monograph on the experimental health plans in 1945. In a letter to the secretary of agriculture, Senator Pepper underscored his committee's interest in the FSA model, highlighting the agency's use of public subsidies and the credible partnership the agency had built among farmers, the government, and the medical profession.

> We believe that these plans, all based on the tax-assisted voluntary health-association principle, constitute a series of experiments of interest to the whole Nation. They are particularly important at this time, when our whole future national health policy is being decided. The experimental plans undoubtedly offer in practice a test for the ideas of those who consider that tax-assisted voluntary health associations might be the solution to this problem.[21]

The FSA programs were "valuable not only to those who are considering starting prepayment plans," Pepper added, "but also to those who are weighing their attitudes toward compulsory health insurance."

In a brief foreword to the monograph, the FSA's chief medical officer echoed Pepper's comments, and commented on the significance of the FSA's experiment in delivery of medical care to the entire nation. "The implications of this report go far beyond those bearing on day-to-day administration," wrote Mott, "It holds lessons for every rural community and for urban as well as rural America. At a time when the Nation is facing its health problems, weighing possible solutions, and rapidly approaching the stage of long-needed action, it is gratifying that this report, with its sound conclusions, is being made available for widespread review."[22]

From these excerpts, it might appear that Pepper and Mott believed that federally subsidized voluntary health insurance plans might offer a solution to the problems of health care access and affordability. This was not entirely the case. The FSA's promotion of a public-private partnership, and the agency's use of tax-assisted financing, were points of real interest. However, the report makes repeated mention of the problems the FSA experienced with enrollment and membership turnover, and linked these directly to the voluntary nature of the agency's programs.

Mott emphasized this point in public comments he made throughout this period. While the 1945 Wagner-Murray-Dingell bill languished in Congress, Mott's "hard-boiled" assessment of the FSA medical care programs was excerpted in the *Journal of the American Medical Association*. He noted the positive impact of the programs on loan repayment and the exposure they gave farmers and physicians to prepaid group health insurance. Mott also emphasized the agency's frustration with membership attrition, adverse risk selection, and rising costs. Mott told the Senate that these problems were interrelated, and that they highlighted the inherent flaws of voluntary approaches to health insurance.[23]

Up to this point, Mott had said little that had not already been written about the FSA programs. It was his concluding remarks, however, that caught the attention of organized medicine and that were quickly disseminated in medical journals around the country. "*We see no answers,*" said the FSA chief medical officer, "*except a more or less universal program for the whole population.*"[24] Mott's comments reached a medical community already primed to do battle against a bill that one congressman called "the most socialistic measure that this Congress has ever had before it."[25]

The AMA marshaled the full weight of organized medicine against national health insurance following the introduction of the 1945 Wagner-Murray-Dingell bill. With the aid of an aggressive and well-financed public relations campaign, the AMA launched a two-pronged assault on national health insurance and its supporters. One line of attack portrayed the idea as socialistic and foreign, a proven tactic that was used with devastating effectiveness in the charged political atmosphere of postwar America.[26] The AMA's second offensive against national health insurance was grounded in the advice of the AMA's media consultants that "you can't beat something with nothing."[27] Voluntary health insurance was the "something" organized medicine proffered as a uniquely American alternative to government-controlled medicine. Notwithstanding the modest penetration of third-party insurance into rural communities, the AMA leadership placed great emphasis on developing a rural strategy as part of its effort to defeat national health insurance, an affirmation of the seriousness with which the medical community still viewed the government's interest in rural medical care delivery.[28]

It is tempting to attribute physicians' motives for opposing compulsory national health insurance to economic self-interest. Certainly, many physicians and organized medical groups suspected that compulsory plans would result in unrestricted demand for services, lower incomes for physicians, and a loss of professional autonomy and control.[29] However, this explanation fails to capture the real concern that many doctors felt about providing the best possible medical care to their patients. Physicians and local and state medical groups were not motivated by venal considerations alone—as the history of the FSA medical care program demonstrates. Nor were they always and invariably in lockstep with the AMA on the issue of national health insurance. As noted in Chapter 2, several mainstream medical groups broke with the AMA, albeit briefly, during the 1930s by supporting compulsory insurance plans. Polls taken during the 1930s and 1940s also indicated that like it or not, many physicians believed that national health insurance was inevitable, and that many in fact favored it.[30]

On the surface, opinions in the medical community about the role of government in medicine and national health insurance appeared cacophonous. However, the core principles that guided the medical profession in its collective reaction to health care reform were never far from view. These principles included free choice of physician, preference for voluntary and private solutions, and wariness toward federal involvement in medical care. "It should be our firm purpose to preserve for all a high quality physician, the freedom to select their own physician and the privilege of being self-sustaining, if possible," wrote a California physician in 1938. "In all these efforts our present system of free and independent medicine must be preserved against governmental bureaucracy and efforts to subject human suffering to political blunder."[31]

While the motivation of the AMA in opposing national health insurance may be suspect, practicing physicians' concern that it would bureaucratize medicine and drive a wedge between doctors and their patients was genuinely felt. Many also believed that national health insurance represented an abdication of professional responsibility. Before the issue of health insurance became moot, state and local medical societies as well as individual doctors commonly argued that the profession and the public alike would be better served if physicians cared for the medically indigent for free, rather than cede control to any third parties. When the doctors were faced with a choice between private or public incursion into health care services, however, the vast majority supported private plans over government-sponsored ones.

The medical profession's influence over the national health insurance debate was contingent to an important degree on the tacit support of the American public. Some scholars have claimed that this support was coerced, while others suggest that it was earned. The fact remains that it was given. Moreover, the medical profession knew this to be the case and used its power to full advantage. Ordinary physicians' and organized medicine's confidence in their ability to persuade the American public that its interests were in line with those of the medical profession grew steadily throughout the 1940s. As one Washington State physician commented: "When push came to shove, 97 percent of the people will support the doctor."[32] In the same vein, a 1947 editorial in the *Journal of the American Medical Association* boasted that "if the family doctor tells John or Mary that Blue Cross is good, then John and Mary and many other John and Marys enroll in the prepayment plans."[33]

Liberal reformers, too, must accept a measure of blame for the failure of the 1945 Wagner-Murray-Dingell bill. Once again, interagency jealousies, personality clashes, and genuine differences of opinion, combined with Roosevelt's prior hesitance to commit political capital to national health legislation, had fractured the reform movement and made the task of those opposed to an expanded federal role in health care financing and delivery that much easier.

By the mid-1940s, public support for national health insurance had waned. The results of a 1946 Gallup poll indicated that support for compulsory health insurance stood at less than 12 percent, a turnaround from earlier surveys that appeared to show majority support for universal health insurance. The reasons for the variability found in surveys of the period—from which both advocates and opponents of national health insurance drew support for their views—is a matter of some dispute among historians. Some scholars suggest that these differences reflected how questions on health insurance were framed, others that they indicated the malleability of public opinion.[34] These explanations touch only tangentially on another possibility. Embedded in the national health insurance debate was a deeper tension between what one might call national character traits. On one level, guaranteeing access to medical care resonated with Americans' egalitarian sensibilities. Compulsory schemes, on the other hand, awakened an instinctive defense of individual liberties and aroused the public's apprehension about centralized government authority.

Historically, opponents of national health insurance have achieved their success by couching their misgivings in terms of threatened freedoms, such as losing the freedom to choose one's own doctor, a physician's freedom to practice without restraint, or the public's freedom from paying more in taxes to fund such a program. Had the Social Security Act been proposed after the national economic recovery had begun, rather than during the bleakest years of the Great Depression, one is left to wonder whether it too might never have been enacted.[35]

A MOST DISCOURAGING PERIOD:
THE DEMISE OF THE FSA MEDICAL PROGRAMS

The end of World War II augured a period of escalating public and political anxiety over the rise of world Communism. Although the Red baiting that characterized American politics reached its zenith in the McCarthyism of the 1950s, it found expression in American medicine in the waning years of the FSA medical care programs. Inspired by the AMA, attacks against reformers within the medical profession and others involved in the national health insurance debate grew more malignant. Although the accusations made against proponents of national health insurance had a conspiratorial tone before the end of the war, the atmosphere became far more poisonous in the postwar period. "American communism holds this program as a cardinal point in its objectives," warned *California and Western Medicine*. "In some instances, known Communists and fellow travelers within the Federal agencies are at work diligently with Federal funds in furtherance of the Moscow party line in this regard."[36] The widespread belief that the federal government had been infiltrated by Communists fueled a backlash against liberals and progressives in the government, and led directly to purges of various federal agencies, including the FSA Health Services Branch. More than a few individuals were driven from public service, seeking calmer waters in academia or in international relief and health-related efforts.

The AMA reserved a special place in its pantheon of pariahs for I. S. Falk and the medical economist Michael M. Davis. In 1947, the *Journal of the Arkansas Medical Association* referred to national health insurance as "the brainchild of such radical social reformers as Davis and Isadore Falk who would like to see our democratic system transformed into a totalitarian system."[37] An AMA-inspired 1949 poster entitled "The House of Falk and Davis" placed the two reformers at the center of the conspiracy to nationalize medicine.[38]

Joining Falk and Davis in the sights of the AMA were a number of high-profile government physicians, including Thomas Parran and Martha May Eliot. In spite of their lukewarm support for the Wagner-Murray-Dingell legislation, both Parran and Eliot were publicly censured by the AMA for their liberal views on health insurance and other matters. In 1946, for example, the AMA passed a resolution denouncing Parran for political activities that were "in opposition to American democratic processes."[39] One year later, the AMA claimed to have "documentary evidence" linking the surgeon general and other government officials to "movements for compulsory health insurance in other countries." Following a series of national meetings on health care sponsored by the U.S. Public Health Service, the AMA accused Parran and other "authors of propaganda" of misappropriating federal funds and "stumping" for national health insurance. This accusation derived in part from Parran's decision to dispatch a Public Health Service physician on a fact-finding study of New Zealand's national health care system. The medical press severely criticized Parran for his decision, some of which was infused with anti-Semitic allusions to the medical officer in question. The unfortunate doctor was investigated by the FBI and later summoned before the House Un-American Activities Committee.[40] Also included in this group of suspect government physicians were the FSA's Fred Mott and Milton Roemer, as well as Assistant Surgeon General Joseph Mountin and George St. John Perrott, two career medical officers whose participation on the Technical Committee for Medical Care had also not gone unnoticed by the AMA. With all the controversy surrounding Parran, it was no surprise that Truman refused to reappoint him as surgeon general in his second term. Truman's decision effectively ended the U.S. Public Health Service's flirtation with a national health program. Parran's immediate successors steered the agency back into its more traditional and politically neutral role.

The AMA's list of conspirators also included social reformers and medical activists who worked outside of the federal government. This group included academics such as Henry Sigerist, director of the Institute of the History of Medicine at Johns Hopkins, and Nathan Sinai, professor at the University of Michigan School of Public Health. All were castigated for their outspoken support for reform in the nation's health care system. [41]

Fred Mott and Milton Roemer of the FSA's Health Services staff in Washington experienced harsh censure from organized medicine and its allies. [42] In the late winter of 1946, Mott decided it was time to leave both the FSA and the United States. The provincial government had already offered Mott a position with the Health Services Planning Commission in Saskatchewan, a publicly funded hospital insurance program that became the model on which Canada built its national health system. With some misgivings, Mott accepted the post. In a March 1946 letter to Saskatchewan Premier Tommy Douglas, Mott summarized the difficult circumstances facing the FSA at the time.

Mott also noted that reformers expected him to line up support for the Wagner-Murray-Dingell bill among farmers, and that his departure would disappoint those who counted on his continued advocacy for rural health issues in what he called the "crucial years ahead."[43] His decision was complicated still further by the knowledge that Parran planned to recommend the creation of a permanent chief medical officer position in the Department of Agriculture and that he wanted Mott for the job.

Before leaving the agency, Mott wrote a farewell memorandum to his staff. In it, he expressed cautious optimism about the prospects for the continuation of the medical care programs. However, he could not disguise his disappointment over the failure to enact national health insurance in his home country. "There seems to be real prospect of basic legislation that will place the whole program in a stronger position. I hope our former strength will be regained or surpassed . . . I hope what Canada does will help out here and I look forward eagerly to plunging back into the fight with you after three years."[44]

In fact, the FSA was facing much larger problems by the time Mott left for Canada. FSA leaders had fought a rearguard defense of the agency's programs for years, and the agency had barely avoided being dismantled several times. In April 1946, however, Congress finally passed the third version of the Farmers Home Administration Act introduced by Representative Harold Cooley. Truman signed the bill into law on August 14. According to the terms of the legislation, the FSA was officially terminated on December 31, 1946. Any residual "functions, powers, and duties" of the FSA were transferred formally to the FHA. The new law eliminated nearly all of the federal rural rehabilitation programs established by the Bankhead-Jones Act in 1937. The few programs that survived, including a much scaled-down tenant purchase program, were placed under the control of the FHA.[45] The Health Services Branch continued as part of the FHA for several more months while the medical care cooperatives, migrant health program, and the experimental health plans were liquidated, or in a few instances transferred to local agencies.[46]

The 1946 midterm election represented the final repudiation of liberalism and the New Deal. The election ushered in a decidedly conservative Congress and returned the Republican Party to power. If reformers saw their chances to enact national health insurance dim as bill after bill died in Congress, the 1946 election quashed any pretense of hope to which they still clung. Although President Truman delivered a strongly worded speech to Congress on behalf of national health insurance soon after the election, the second Wagner-Murray-Dingell bill expired in committee in the Republican-controlled Congress the following spring.[47]

In the early winter of 1946, a former colleague of Fred Mott wrote to his friend, who was by then safely ensconced in Canada. Jesse Yaukey was a U.S.

Public Health Service statistician who had worked closely with the FSA, and his dispirited letter captured the feeling of many liberals who still remained in government service in the aftermath of the November elections. "The significance of the outcome of the November 5 election is very real for us," wrote Yaukey, "both in terms of the Congress it is giving us and in the trend of thinking it indicates . . . The tide is running pretty strongly against us." Whereas in previous years, the FSA medical care programs had often been criticized as part of the broadsides against the agency itself, this was no longer the case. Yaukey stated that the Health Services Branch was now the "immediate point of attack" for the agency's opponents and that its already skeletal staff had been purged recently in order to "make the organization acceptable."[48] Yaukey's letter to Mott also indicates that he and others in government service feared political repercussions as a result of their views, and the liberal exodus from government gained momentum in the wake of the election.

Mott shared Yaukey's worries about the direction that national politics had taken in his native country. In a letter written to his former colleague a few months later, Mott provided an appropriate epitaph to the FSA medical care program in which he had played such a key role: "You have all been going through a most discouraging period. There has been a rather ominous silence for several weeks now, I only hope that the program is not being wiped out altogether . . . Perhaps our consolation can be that we played a crucial role at a time when pioneering was urgently needed. A great deal could still be done, but it seems pretty hopeless in view of the present character of the agency and of current trends in the US."[49]

Despite the results of the 1946 midterm elections, national health bills were introduced in Congress in 1947 and again in 1949. Truman made national health insurance a campaign issue in the 1948 presidential election, hoping to prevent liberal Democrats from defecting to the Progressive Party candidate, former Secretary of Agriculture Henry Wallace. Following Truman's astonishing come-from-behind victory over Thomas Dewey, organized medicine steeled itself for a battle that never materialized. However, Congress showed little interest in any health insurance proposals, and none came close to passage.

The FSA cast a slowly fading shadow over the national health insurance debate before finally disappearing from view. It is a measure of organized medicine's concern over the FSA medical care programs, and the role that FSA medical leaders played in the national health insurance movement, that attacks against the FSA continued even after the agency was legislated out of existence. "The drive to federalize medicine, the entering wedge, and other fields of human endeavor," editorialized the *Journal of the Ohio State Medical Society* in May 1947, "stems from a well-organized group, part of whom hold positions in Federal government; and has recently been concentrated on the farmers."[50]

The coup de grâce for the national health insurance movement was delivered, courtesy of the AMA, in the 1950 midterm elections. Assisted once again by a savvy public relations firm and a war chest swollen with mandatory contributions from its membership, the AMA mounted a highly effective political campaign and defeated nearly every congressional candidate who refused to swear off support for national health insurance. It is telling that after nearly two decades of emotionally charged national debate, Congress never took a formal floor vote for or against any of the national health insurance proposals.[51]

CHAPTER SEVEN

Conclusion

I know it has been discouraging and piecemeal, but you have laid the foundation on which a real program will ultimately be built.
WILL ALEXANDER TO FRED MOTT, 1945

The FSA was a tangible manifestation of the New Deal's desire to improve the economic well-being of the nation's most impoverished rural citizens during the Great Depression.[1] Working with funding that was totally inadequate for the size of the economic crisis, the agency materially improved the lives of hundreds of thousands of rural families and raised the nation's awareness of the extent and severity of rural poverty. FSA leaders pursued their mission with a creativity and purposefulness that mark the agency as a singular experiment in federal social welfare policy. Although the FSA had neither the resources nor the political support to seriously attack the causes of rural poverty, the agency challenged the traditional authority of the agricultural establishment in rural America. For this reason, the agency incurred the wrath of powerful farm organizations and their political supporters and became a favorite whipping post for conservative opponents of the New Deal.

The FSA's medical care programs were among the agency's most imaginative and influential accomplishments. They were also the New Deal's largest organized experiment in health care delivery. The programs' broad coverage; their national reach; and their impact on rural farmers, rural physicians, and national health politics make their history a dramatic episode in the history of twentieth-century American medicine and federal health policy. The opportunity to conceptualize, organize, and administer a national medical care delivery program also honed the skills that many who worked in the FSA medical care programs brought to their future careers in academia, public health, and medical care. My primary aim in writing this book has been to give this experiment in government-sponsored health care delivery its historical due.

The FSA medical care programs ameliorated the physical suffering of hundreds of thousands of indigent farm families and provided many of them with

Hoe culture, Alabama, 1936. Vicinity of Anniston, Ala., June 1936. Dorothea Lange.

their first taste of regular medical care, including ordinary physicians' services, hospitalization, preventive services, and health education. The FSA based nearly all of its plans on the group prepayment principle at a time when third-party insurance was just beginning to gain widespread acceptance. The attention the FSA medical care programs received from organized medicine and farm groups, and the agency's role in nurturing the awareness and acceptance of prepaid group insurance in agricultural communities, pushed rural health

Tubercular wife and daughter. Nipomo Valley, Calif., 1936. Dorothea Lange.

issues to the center of national health policy and politics for the better part of a decade.

The FSA also experimented on a national level with concepts that are salient features of our current health care system, including capitation, peer review, and an emphasis on promotion of health and prevention of disease. Physicians were essential participants, but they were but one part of a multi-faceted approach to health care that included the services of nurses, dentists, health educators, sanitarians, and home management supervisors.

Although the FSA's medical leaders borrowed freely from concepts circulating at the time, they field-tested these ideas on a scale unmatched by other medical care programs of that era. In doing so, the FSA extended the boundaries of the relationship between the government and the medical profession at a time when fundamental changes in that relationship seemed very much within reach. Ironically, historical commentary on the Farm Security Administration and its medical staff's role in the national health insurance movement has barely been noted, or dismissed altogether. At the time, however, supporters and opponents alike viewed the FSA leadership and the agency's medical care programs as very much a part of the national health insurance movement. The New Deal social welfare system, whose cornerstone was Social Security, was never extended to cover the health care of all of the nation's citizens. Still, the FSA's incursion into the practice of medicine and the organization of medical services established important precedents on which later federal health care initiatives were built, especially those focused on disadvantaged communities. For example, the FSA's multidisciplinary model was resurrected in the community health center and health maintenance movements of the 1960s and early 1970s.

How did the FSA medical leaders achieve what they did? After all, the FSA developed its plans at a time when federal involvement in medical care was an issue that divided politicians, the medical community, and society in general. FSA administrators defended the medical care programs from critics by portraying them as a modest effort to improve the access of low-income farmers to medical care in order to facilitate their economic rehabilitation and ensure the repayment of government loans. The FSA also established a serviceable partnership with the medical profession in rural communities. This partnership included not just physicians and the government, but to an important degree indigent farmers themselves. FSA medical leaders made the programs palatable to rural physicians and state and county organized medical groups by focusing primarily on low-income families and by accommodating the most cherished values of the medical profession, including voluntary participation, a significant measure of professional autonomy, and free choice of physician. Finally, the economic advantages to rural physicians were significant. The FSA programs stabilized the declining incomes of rural physicians and bolstered their sagging professional status at a time when the locus of medical care in the United States was shifting to hospitals and specialists.

The FSA's decentralized approach to rural rehabilitation lent itself to policies that encouraged FSA clients and local interests to influence the agency's programs. FSA leaders portrayed this flexibility as a programmatic virtue that encouraged participatory democracy and fostered program expansion. This was certainly true. At the same time, however, even the agency staff and those critics sympathetic to the FSA's mission acknowledged that the agency was sometimes too solicitous of entrenched power groups. This stance has been the cornerstone for much historical criticism leveled at the FSA.

The FSA medical care programs suffered from the same difficulty. Negotiations between the FSA and organized medical groups consumed a great deal of time and effort. Maintaining the agency's programs also required constant vigilance on the part of the FSA's Health Services personnel. In addition, local physicians and medical societies could choose not to participate, could desert the plan, or could (and did) sway the opinion of families and physicians toward the FSA plans. Although physicians did not dictate the organization or the operation of the FSA medical care programs, they wielded considerable influence over the success or failure of any given medical care unit.

AMA leaders distrusted the motivations of the FSA and they were obstructionist. However, they could not completely ignore the economic pressures on rural physicians or the inadequate health status of indigent farm families. Consequently, the AMA usually deferred to state and county medical societies, a significant number of whom supported the FSA programs. This deference ended only when the FSA leaders aligned themselves openly with the national health insurance movement in the early 1940s, a decision that undermined support for the agency among its erstwhile physician allies as well.

The FSA's decision to advocate national health insurance came at a time when the agency was already in decline. Waning support for the New Deal's social agenda, growing congressional conservatism, changes in demography and national priorities caused by World War II, and an improving national economy all played a role in the difficulties the FSA experienced. The FSA tried to adapt to changing social, economic, and political circumstances during the war. However, the agency's supporters could not match the financial and political clout of its enemies, and they could not prevent the attrition of the agency's programs, staff, and mission. By the time Congress legislated the agency out of existence in 1946, the FSA was a shell of its former self.

The emergence of voluntary health insurance and the alliance between conservative farm groups and organized medicine in its support further weakened the FSA medical care programs. Before the Great Depression, medical care was by and large a simple cash transaction between physicians and patients. As physicians and the American public embraced voluntary health insurance, organized medicine's arguments against federal involvement in medical care delivery and financing grew more insistent and convincing. By the end of World War II, the federal government's role had crystallized into one largely directed at funding for medical research, education, public health, and infrastructure building. The long debate as to whether expanding access to health care was a public responsibility or a professional duty best managed through private means was settled in favor of the latter. Not until the 1960s would there again be sufficient political consensus to broaden the federal government's responsibilities to include improving access to medical care for selected disadvantaged groups.[2]

The coalition between the FSA and rural physicians had always been a delicate one. Once it was damaged by the FSA's support of national health insurance, forces operating on a national level simply overwhelmed whatever residual support the agency and its programs still had with rural doctors. In the political climate of the postwar period, these developments had damaging professional and personal repercussions for those who worked in the FSA Health Services Branch, as well as for those who were publicly aligned with the national health insurance movement.[3]

The obstacles facing the FSA at the time made the decision to advocate national health insurance understandable, but the character of both the agency and its Health Services staff made it inevitable. Publicly, FSA leaders rationalized their involvement in medical care on economic grounds, and there is evidence that the agency's programs improved the physical and economic status of FSA clients. However, the comprehensiveness of the medical care programs and the alliance of FSA leaders with kindred spirits in government who formulated and directed national health policy speaks to a more expansive and not wholly reactive philosophy.

The FSA attracted individuals who believed that restructuring the medical care system was necessary in order to provide medical care to all the nation's citizens, especially its most disadvantaged. They were confident that the FSA's medical care programs offered valuable lessons for federal health planning. The attention the agency's medical programs received from those involved in federal health policy, from the media, and from individuals and organizations on both sides of the national health insurance debate reinforced their confidence.

The FSA medical care programs never became the nucleus from which comprehensive reform of the nation's health care system evolved. However, the agency was drawn into the debate on the government's responsibility to promote health and ensure access to medical care for all Americans during a period when the passage of national health insurance seemed within reach.

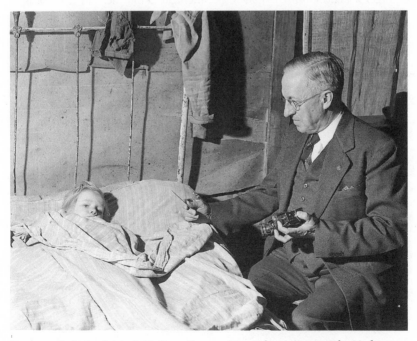

Doctor at bedside of sick child. Scott County, Mo., February 1942. John Vachon.

The national health insurance movement, and the feints that the federal government made into health care, helped direct the evolution of the health care system in the United States into its present form. The role of the FSA and its medical leaders in the national health insurance debate, using rural communities as a "dry run," makes its history all the more valuable to those interested in health care reform in the United States and in understanding how we obtained the system of health care that we now have.

There are rare moments of convergence in history when the lives of individuals are touched in concrete and lasting ways by broad social movements, and when personal commitment merges with a sense of collective responsibility. The Great Depression and the New Deal was such a time. The crisis of the 1930s tested the nation's sense of purpose and forced it to grapple with long-standing social inequalities. Although historical opinion about this period of our history is plentiful and diverse, few dispute that the federal government became a palpable presence in the lives of millions of ordinary Americans, or that the New Deal was a beacon for a host of men and women who believed in government activism to remedy social ills. Whatever the New Deal's shortcomings, the individuals whose remembrances of it and of the FSA have enlivened this history accept that the measure of an idea—or for that matter, of a federal program—goes beyond data and dollars. They felt part of a larger social movement—an "experiment in democracy" in the words of T. G. Moore—and its influence was a lasting one.

Epilogue
The Legacy of the FSA Medical Care Programs

My own experience has been that when you go into an area to heighten the consciousness of the people to what could be done, and do the basic kind of education, together they can move on by themselves. You serve as a catalyst and then the thing begins to unfold.
THOMAS GARLAND MOORE JR., 1979

The legacy of the Farm Security Administration is a rich one. In the postwar era, for example, developing nations in Central America, South America, and Africa benefited from the rural rehabilitation expertise of former FSA staff, and many adopted the agency's model of cooperative farming.[1] The concepts, strategies, and philosophy that characterized the FSA's rural medical care programs also found expression in later years in health care delivery programs in the United States and abroad. Former FSA Health Services Branch staff were especially influential in public health and medical care circles. A number of individuals with close ties to the FSA have been important figures in the American Public Health Association. At least five became professors in medical schools or in schools of public health, including Fred Mott at the University of Toronto; Milton Roemer at UCLA; George Silver at Yale; Leslie Falk at Meharry Medical College in Nashville, Tennessee; and Solomon Axelrod at the University of Michigan. Thomas Parran, who as surgeon general supported the FSA's programs, later helped create both the World Health Organization and the Pan-American Health Organization. John Newdorp, who later worked alongside Mott with the United Mine Workers of America (UMWA), summed it up as follows: "Just look at some of the names of the people and see where they have gone subsequently . . . That is the kind of contribution that the program made. All of these people had contact with these programs and had further positions in the field of medical care administration or public health . . . We had a bunch of them in the FSA."[2] This epilogue explores just a few of these linkages.

MOTT HEADS NORTH: THE FSA AND
THE CANADIAN NATIONAL HEALTH SYSTEM

In 1944, the Social Democratic Party swept into power in Saskatchewan, Canada. The Social Democrats supported universal health insurance, and after the election the newly elected premier, Tommy Douglas, invited Henry Sigerist of Johns Hopkins University to come to the region to assess the state of health and medical care. Sigerist suggested creating a series of regional health services commissions that would use local taxes to pay the costs of hospitalization. Sigerist's recommendations provided the conceptual framework for what eventually became the province's hospital insurance system, which was the first program of its kind in Canada. The program was centrally administered and financed through tax revenues. A quasi-governmental body, the Health Services Planning Commission, supervised the insurance plan. All of Canada's provinces later adopted this model, earning Saskatchewan the right to consider itself the birthplace of the Canadian health care system.[3]

The system's ties to the FSA began in early 1946 when Douglas asked Fred Mott to chair the Health Services Planning Commission.[4] Saskatchewan's economic base was agriculture, and like their neighbors south of the border, the region's farmers suffered terrible economic hardship and an erosion of their health status during the depression of the 1930s. Mott's liberal views on the role of government in medical care delivery, his administrative experience, and his familiarity with rural health issues and rural communities were outstanding qualifications that made him a masterful choice. Mott succeeded Dr. Cecil Sheps in the post. Sheps was Canadian and an influential figure in provincial political circles. However, he antagonized the province's medical community, a problem that the new government could ill afford. Mott used his prior experience working with physicians and organized medical groups to great advantage in Saskatchewan, and his five-year tenure with the Health Services Planning Commission was marked by cordial relations with the province's medical community.[5]

The similarities between the FSA medical care programs and the health insurance system that evolved in Canada are striking. Both programs used public financing and central administration through a third party that was at least nominally separate from the government. In the case of Saskatchewan, this was the Health Services Planning Commission. Like the FSA, the provincial government negotiated with organized medical groups on organizational, administrative, and fiscal matters. As a direct result, the Saskatchewan plan also preserved freedom of choice for patients and modified but did not abandon fee-for-service practice.[6]

Mott later indirectly acknowledged the FSA programs in a 1948 *Canadian Medical Association Journal* article in which he discussed the government's plan to expand the provincial insurance plan to include ordinary physicians' services as well as hospital care. Although this expansion did not occur until 1961, Mott stated that the Health Services Planning Commission planned to take "full advantage of the experience that has been accumulated elsewhere in formulating our plans for an effective means of meeting the great and growing demand for general medical services."[7]

In addition to his duties with the Health Services Planning Commission in Saskatchewan, Mott was Acting Deputy Minister of Public Health from 1949 to 1951, and served as Deputy Minister in 1951. He also represented Canada as a delegate to the fledgling World Health Organization, a rare honor, given his nationality (Andrew Mott to author, personal communication). Mott remained a steadying influence on Saskatchewan's health care system until 1951, when he returned to the United States to take a position with the United Mine Workers of America.

Mott was not the only former FSA staffer to work in Saskatchewan. Milton Roemer worked in the provincial health department for three years beginning in 1953 before he too returned to his native country. It is ironic that Sigerist, Mott, and Roemer helped Canada accomplish what they and their colleagues had fought for, but could not achieve in the United States.

BUILDING ON THESE EXPERIENCES: THE FSA AND THE UNITED MINE WORKERS OF AMERICA

Health and safety were central negotiating issues during the national coal strike in 1946. John L. Lewis, who was the president of the United Mine Workers of America, eventually extracted an agreement from the coal owners that helped the UMWA build one of the largest and most comprehensive union-sponsored medical care delivery programs in the United States of the postwar era, the UMWA Welfare and Retirement Fund. In 1946, three physicians with extensive FSA experience—Lorin Kerr, John Newdorp, and Allen Koplin—met with other UMWA staff on the outskirts of Washington, D.C., to craft the proposal Lewis used in his negotiations with the coal operators.

In its earliest form, the UMWA Welfare and Retirement Fund was an indemnity plan paid for through employee payroll deductions and contributions from coal owners that was based on a percentage of the tonnage mined by union members. Consistent with postwar trends in health insurance, the UMWA initially focused on hospitalization coverage. The high trauma rate among miners led the union to include rehabilitation services as well. In time, the UMWA expanded the program to include comprehensive health benefits to miners and their families through a network of medical clinics and hospitals.[8] In fact, it was the offer to supervise the union's hospital program that led Fred Mott to leave Canada in 1951.

No postwar health care delivery program in the United States had a more direct connection to the FSA than the UMWA programs. As many as twenty FSA Health Services staff members worked for the UMWA at some point in their career. In 1947, Leslie Falk was recruited to be medical director of a UMWA health clinic outside of Pittsburgh, Pennsylvania. Falk stayed with the UMWA for over twenty years, and was well known in the Pittsburgh area for his commitment to occupational health and safety issues, civil rights, and other social causes. From 1951 to 1958, Fred Mott was the medical director of the Miner's Memorial Hospital Association. John Newdorp was his assistant.[9] Other FSA staff members who participated in the UMWA programs included Ken Pohlman, who had been chief of the FSA's division of Health Services specialists; Henry Daniels, a hospital administrator who supervised the agricultural workers health associations from Washington at the end of the war; and Dr. Henry Makover, who succeeded Mott as chief medical officer for the FSA and WFA, and who briefly held the post intended for Mott in the Department of Agriculture.[10]

Lorin Kerr stayed with the UMWA for his entire career, eventually becoming the first physician in the United States to be named director of a union health and safety program. From his position with the UMWA, Kerr played an instrumental role in the passage of the 1969 Federal Coal Mine Health and Safety (Black Lung) Act as well as the 1970 Occupational Safety and Health Act. Like many of his former colleagues, Kerr also made his mark on the American Public Health Association (APHA). He was active in APHA's Occupational Safety and Health section and was elected president of the organization in 1974.[11] Kerr spoke for his FSA colleagues when he explained that "the work we did with the migrants became the basis on which we built the UMWA Welfare and Retirement program . . . We had to build on these experiences that we all had."[12]

There were many programmatic resemblances between the FSA medical care programs and the UMWA health programs. For example, the union's decentralized network of clinics allowed for a measure of local flexibility, and the participation of union members in the program was encouraged. Like the FSA, the UMWA program was centrally administered and funded. Benefits for miners and their families were also comprehensive and marked a significant improvement in their access to medical care that was free from the control of coal operators and company doctors.

Full-time salaried physicians were used to a much greater degree in the UMWA programs than they had been in the FSA plans. The UMWA prized former military physicians because they were well trained, and as a consequence of their military experience they were less opposed to salaried practice. The use of salaried physicians and the construction of its own health care facilities enabled the UMWA to build its program in spite of the hostility it faced from local and state medical societies.

Although the FSA's connections to the UMWA are especially significant, former FSA staff had ties to other union-sponsored experiments in health care delivery. In 1957, for example, Walter Reuther of the United Auto Workers recruited Mott to be the executive director of the Community Health Association, a consumer-controlled group prepayment program sponsored by the United Auto Workers union for its members and their families in the Detroit metropolitan area.[13]

POVERTY WARRIORS REDUX:
THE FSA AND THE WAR ON POVERTY

The debate over whether improving access to medical care was a public responsibility or a professional duty simmered following the demise of the national health insurance movement in the late 1940s. It resurfaced as a major social and political concern during the War on Poverty in the 1960s. Like the New Deal, the War on Poverty used social insurance and selected rehabilitation to improve the living conditions of those bypassed by the unequaled economic prosperity of postwar America. The War on Poverty targeted a wide range of societal ills, including the substandard health status and the inadequate access to medical care of the nation's low-income citizens.

In 1965 Congress passed three amendments to the Social Security Act, creating Medicare and Medicaid. Congress's action—over the objections of a coalition that looked very similar to the one that defeated national health insurance during the 1940s—effectively ended more than two decades of controversy on whether and how to extend publicly funded health insurance to the elderly and the poor.[14] Although they were passed as a compromise legislative package, the two programs have important differences that in large part have determined the political fortunes of both in subsequent years. Part A of the Medicare program provides hospital insurance to those eligible for Social Security and is financed by a system of mandatory payroll deductions. Part B of the Medicare program is a voluntary insurance program that covers physicians' services for eligible retirees. Medicaid provides health insurance to the poor through a program of federal grants to the individual states. Eligibility and specific coverage are determined largely by the states.

If the War on Poverty was reminiscent of the New Deal, then the Office of Economic Opportunity (OEO) has the closest claim of kinship to the Farm Security Administration. Congress created the OEO in 1964, and the agency soon became one of the most visible expressions of a renewed federal commitment to strike at the sources of poverty. The OEO earned a reputation as the guardian of the poor and politically disenfranchised, a mantle once worn by the FSA. Although the OEO had an overwhelmingly urban emphasis, later amendments increased funding for rural communities.[15]

The Farm Security Administration and the Office of Economic Opportunity shared many similarities in their broad rehabilitation missions, the strategies each agency used to achieve their objectives, and their reform rhetoric. For example, OEO leaders believed that promoting local leadership and inculcating democratic values in poor communities would advance their larger goal of social change and community renewal. Both agencies depended on a large and dedicated field staff to pursue their rehabilitation mission. Finally, while health care was not part of the OEO's enabling legislation, the agency soon learned that it was a concern in many communities, and it was not long before community clinics and neighborhood health centers were springing up with the OEO's blessing and financial support.[16]

President Johnson's domestic advisors had to be familiar with earlier federal rehabilitation efforts, including those sponsored by the FSA. No fewer than eight participants in the 1965 White House Conference on Health had once worked for the agency. In addition, Dr. George Silver, who had supervised the farm labor programs in the Northeast during the war, was named deputy assistant secretary in the Department of Health, Education and Welfare during the Johnson administration.[17]

There were individuals who worked for the Farm Security Administration and later the Office of Economic Opportunity as well. In 1967, for example, the OEO helped recruit Leslie Falk to the Department of Preventive and Family Medicine at Meharry Medical College. From this position, Falk supervised an OEO-funded neighborhood health center in Nashville, and oversaw the agency's regional network of community health centers. In his sociological history of federal health programs for African Americans in the South, Couto writes that Falk was chosen because the OEO's regard for him "as an 'old-FSA' type made him valuable enough to work to relocate him and match him with new initiatives like those which had worked before."[18]

Both the FSA and the OEO built bridges between the federal government and the medical community, in the latter case primarily through affiliations with teaching hospitals and medical schools. The OEO viewed traditional institutional relationships among hospitals, medical schools, and community health centers as barriers that prevented a segment of the community from receiving needed medical care. Because the agency used its funding to reconfigure these long-standing relationships—for example, by fostering community-based leadership in running health centers—the agency's efforts engendered much controversy. This contrasts with Medicare and Medicaid, two programs that were marketed as publicly funded health insurance plans for disadvantaged citizens and were never intended to reorganize the delivery of medical care. Although Medicare and Medicaid were not received warmly by organized medicine at the time, their incrementalism was a strong point in their favor with the American public and the medical community at large.

OEO leaders encountered obstacles in developing health care programs that would have been familiar to their counterparts in the FSA days. Conflict between the government, the community, and medical providers over who should control the community health centers was common, for example. In addition, because the OEO targeted African American communities, the agency's presence in the South was considerable. Predictably, this embroiled the OEO in the racial politics of the region.

Like the FSA health programs, OEO-sponsored community health centers frequently ignored narrow definitions of health care by offering health education, nutrition, and sanitation programs along with medical services. The efforts of both agencies to intertwine traditional public health concerns with a medical care delivery program ran counter to the historical trend that had seen clinical medicine and public health drift apart for decades. One of the earliest OEO-sponsored rural clinics, the Tufts Delta Health Center in Mississippi, epitomized this approach. The health center provided ordinary medical care, promoted local farming cooperatives, and supplied technical advice and financial support to improve sanitation, sewers, and housing. Although the Kark's multidisciplinary health center model in South Africa is often mentioned as the archetype for the OEO's community health centers, this brief description suggests that more indigenous roots exist as well. Indeed, a number of rural health centers sponsored by the OEO in the South were established in communities that were once home to FSA medical care plans.[19]

The OEO's emphasis on indigent communities also left the agency more vulnerable to changes in the political climate. When the Vietnam War shifted the nation's attention from domestic to geopolitical concerns, Congress narrowed the OEO's mission and curtailed its budget and programs. In this regard, the OEO's experience was similar to that of the FSA during World War II. This contrasts with programs, including Social Security and Medicare, that have a large middle-class constituency. Such programs have proven more sustainable politically and more amenable to involvement of the private sector, whereas the fortunes of programs that target the poor—such as the FSA, the OEO, and Medicaid—have risen and fallen according to the strength of the federal government's advocacy.

Notes

PREFACE

1. I borrowed this phrase from Richard Couto, whose encouragement for my research on the FSA has been steady for over twenty years. Couto, *Ain't Gonna Let Nobody Turn Me Around: The Pursuit of Racial Justice in the Rural South* (Philadelphia: Temple University Press, 1991).

INTRODUCTION

Epigraph: Telephone interview by author, Sacramento, Calif., 24 November 1979.

1. Thomas Garland Moore Jr., interview by author, Sacramento, Calif., 11–12 February 1984.

2. The reasons for this historical neglect are hard to ascertain, although the FSA itself has languished in relative historical obscurity compared to better known New Deal agencies such as the Public Works Administration, Works Progress Administration, Civilian Conservation Corps, and others. A wonderful primary source on the evolution of New Deal rural rehabilitation programs is Berta Asch and A. R. Mangus, *Farmers on Relief and Rehabilitation* (Washington, D.C.: GPO, 1937). The most substantive institutional history of the FSA is Sidney Baldwin's *Poverty and Politics: The Rise and Decline of the Farm Security Administration* (Chapel Hill: University of North Carolina Press, 1968). Useful secondary sources on the FSA and New Deal agricultural history include Paul E. Mertz, *New Deal Policy and Southern Rural Poverty* (Baton Rouge: Louisiana State University Press, 1987) and Roger Biles, *A New Deal for the American People* (De Kalb: Northern Illinois University Press, 1991). There are a number of superb books about the documentary photographs from the Great Depression, including Roy Emerson Stryker and Nancy Wood, *In This Proud Land: America 1935–1943 as Seen in the FSA Photographs* (Boston: New York Graphic Society, 1973); Milton Meltzer and Dorothea Lange, *A Photographer's Life* (New York: Farrar, Straus, Giroux, 1978); and Arthur Rothstein, *The Depression Years, as Photographed by Arthur Rothstein*

(New York: Dover, 1978). Those interested in the medical photographs of the FSA should consult John D. Stoeckle and George Abbott White's *Plain Pictures of Plain Doctoring: Vernacular Expression in New Deal Medicine and Photography* (Cambridge, Mass.: MIT Press, 1985).

3. USDA, Farm Security Administration, *Group Medical Care for Farmers,* Publication no. 75 (Washington, D.C.: GPO, 1941), 1.

4. Baldwin, *Poverty and Politics,* 286.

5. There are a number of sources on the history of agricultural policy in the United States that touch on the FSA contribution and the controversies surrounding the agency and New Deal rural welfare policy in general. In addition to those cited above, see also Gilbert C. Fite, *Cotton Fields No More: Southern Agriculture, 1865–1980* (Lexington: University Press of Kentucky, 1984); John T. Kirby, *Rural Worlds Lost: The American South, 1920–1960* (Baton Rouge: Louisiana State University Press, 1987); Theodore Saloutos, *The American Farmer and the New Deal* (Ames: Iowa State University Press, 1982); Walter J. Stein, *California and the Dustbowl Migration* (Westport, Conn.: Greenwood Press, 1973).

6. Richard Couto, *Ain't Gonna Let Nobody Turn Me Around: The Pursuit of Racial Justice in the Rural South* (Philadelphia: Temple University Press, 1991), 306–7.

7. Baldwin, *Poverty and Politics,* 193.

8. T. H. Watkins, *The Great Depression: America in the 1930s* (Boston: Little, Brown, 1993), 278.

9. Olaf F. Larson, *Ten Years of Rural Rehabilitation in the United States* (Bombay, India: Association of Advertisers and Printers, 1950), 90; Carl T. Schmidt, *American Farmers in the World Crisis* (New York: Oxford University Press, 1941), 232. The FSA's role in the campaign to eradicate pellagra is briefly noted by Elizabeth Etheridge in *The Butterfly Caste: A Social History of Pellagra in the South* (Westport, Conn.: Greenwood Press, 1972), 205. Any serious analysis of the FSA health programs would start with the 1948 book written by two key figures in the FSA Health Services Division, Frederick D. Mott and Milton I. Roemer, *Rural Health and Medical Care* (New York: McGraw-Hill, 1948). Selected passages throughout the text have been published in a series of articles on the FSA written by the author between 1989 and 1994. Permission was granted by each journal for inclusion of these excerpts. See Michael R. Grey, "The Medical Care Programs of the Farm Security Administration, 1932 through 1947: A Rehearsal for National Health Insurance?" *American Journal of Public Health* 84 (10): 1678–87, 1994; Michael R. Grey, "Dustbowls, Disease, and the New Deal: The Farm Security Administration Migrant Health Programs, 1935–1947," *Journal of the History of Medicine and Allied Sciences* 48 (1): 3–39, 1993; Michael R. Grey, "Syphilis and AIDS in Belle Glade, Florida, 1942 and 1992," *Annals of Internal Medicine* 116 (4): 329–34, 1992; Michael R. Grey, "Poverty, Politics, and Health: The Farm Security Administration Medical Care Program, 1935–1946," *Journal of the History of Medicine and Allied Sciences* 44 (3): 320–50, 1989.

Major secondary historical analyses of twentieth-century American medicine that mention the FSA medical care programs are Paul Starr, *The Social Transfor-*

mation of American Medicine: the Rise of a Sovereign Profession and the Making of a Vast Industry (New York: Basic Books, 1982); Rosemary Stevens, *American Medicine and the Public Interest* (New Haven: Yale University Press, 1971); Daniel S. Hirshfield, *The Lost Reform: The Campaign for Compulsory National Health Insurance from 1932 to 1943* (Cambridge, Mass.: Harvard University Press, 1970); and Rickey Hendricks, *A Model for National Health Care: The History of Kaiser-Permanente* (New Brunswick, N. J.: Rutgers University Press, 1993).

10. The institution of quarantines in port cities, for example, was a decision that often involved explicit consideration of the economic impact (pro and con) of the decision. Starr, *Social Transformation,* 240. The U.S. Public Health Service's role in maritime-related commerce is considered in Fitzhugh Mullan's *Plagues and Politics: The Story of the United States Public Health Service* (New York: Basic Books, 1989), 14–31. Mullan built on the work of a previous historical summary of the U.S. Public Health Service by Bess Furman, *A History of the U.S. Public Health Service, 1798–1948* (Washington, D.C.: U.S. Department of Health, Education and Welfare, 1973), 1–13. Economic issues in public health are a common theme in a series of articles reprinted in Ronald L. Numbers and Judith W. Leavitt (eds.), *Sickness and Health in America* (Madison: University of Wisconsin Press, 1985). Two articles by Gerald Markowitz and David Rosner on the activist role of the Division of Labor Standards in the Department of Labor indicate that the FSA was not alone among New Deal bureaucracies taking an ideologically activist stance during the New Deal. This division experienced attacks from the business community that are in many respects analogous to those experienced by the FSA from the farm bloc. Gerald Markowitz and David Rosner, "More than Economism: The Politics of Workers' Safety and Health, 1932–1947," *Milbank Memorial Fund Quarterly* 64 (3): 331–54, 1986; Gerald Markowitz and David Rosner, "Research or Advocacy: Federal Occupational Safety and Health Policies during the New Deal," *Journal of Social History* 18 (3):365–81, 1985.

11. Baldwin, *Poverty and Politics,* 208–9.

12. Richard Hellman, "The Farmers Try Group Medicine," *Harper's Magazine,* December 1940, 5–7.

13. Charles E. Rosenberg, *Explaining Epidemics and Other Studies in the History of Medicine* (Cambridge: Cambridge University Press, 1992), 263.

14. This literature on the history of national health insurance in the United States is too extensive to annotate fully here. Readers interested in an overview of the history of national health insurance should consult any of the excellent histories on twentieth-century American medicine, including Starr, *Social Transformation,* and Stevens, *American Medicine and the Public Interest.* Ronald Numbers's book, *Almost Persuaded* (Baltimore: Johns Hopkins University Press, 1978) is highly recommended for its comprehensiveness and accessibility on this issue as it emerged during the Progressive Era. For a more focused analysis of the national health insurance movement up to the 1943 Wagner-Murray-Dingell bill, see Hirshfield's, *Lost Reform.* Useful analyses written by individuals close to these events are Milton I. Roemer, "I. S. Falk, the Committee on the Costs of Medical Care, and the Drive for National Health Insurance," *American Journal of Public*

Health 75 (8): 841–48, 1985; I. S. Falk, "Proposals for National Health Insurance in the USA: Origins and Evolution, and Some Perceptions for the Future," *Milbank Memorial Fund Quarterly* 55(2): 161–91, 1977; and Arthur J. Viseltear, "Compulsory Health Insurance in California, 1934–1935," *American Journal of Public Health* 61 (10): 2115–26, 1977. The U.S. Public Health Service's role in the national health insurance debate is briefly summarized by Mullan in *Plagues and Politics*.

15. Mott and Roemer, *Rural Health and Medical Care*, 405–6; also, Hirshfield, *Lost Reform*, 86, and Baldwin, *Poverty and Politics*, 208–9.

16. Baldwin, *Poverty and Politics*, 325–35. In a recent biography, Doris Kearns Goodwin describes Eleanor Roosevelt's frustration with her husband's lack of interest in domestic social issues, and her fears that the soul of the New Deal was being lost as a result of his preoccupation with the war: Doris Kearns Goodwin, *No Ordinary Time: Franklin and Eleanor Roosevelt: The Home Front in World War II* (New York: Simon and Schuster, 1994).

17. The FSA's shift in institutional priorities is discussed in Baldwin, *Poverty and Politics*, 325–35. The history of America's internment camps may be found in Michi Weglyn, *Years of Infamy: The Untold Story of America's Concentration Camps* (New York: William Morrow, 1976).

18. The FSA was not the only New Deal agency involved with medicine and health care. The Tennessee Valley Authority set up medical care plans for workers employed on its dam projects. The Works Progress Administration and the Public Works Administration built many hospitals and collaborated with the U.S. Public Health Service on malarial control programs and numerous sanitation projects. Useful primary sources that discuss various New Deal health programs include *Medical Care for Americans* (Philadelphia: Annals of the American Academy of Politics and Social Sciences Press, 1945), 86–167. Hirshfield provides a summary of New Deal agencies' involvement in medical care, breaking these efforts into two groups. One provided medical care as part of a direct relief program, such as the Federal Emergency Relief Administration described in Chapter 1. The other group, in which Hirshfield includes the FSA, the Civilian Works Administration, and the Works Progress Administration, provided medical care as a component of their larger social rehabilitation goals. See Hirshfield, *Lost Reform*, 80–86.

19. The Commonwealth Fund, *Historical Sketch, 1918–1962* (New York: Harkness House, 1963); Julius Rosenwald Fund, *Eight Years' Work in Medical Economics, 1929–1936* (Chicago: Julius Rosenwald Fund, 1937); and Julius Rosenwald Fund, *New Plans of Medical Service* (Chicago: Julius Rosenwald Fund, 1936).

20. Details and statistics in the text follow from Starr, *Social Transformation*, 295–99, 325–34. For specific discussion on the history of Blue Cross, excellent primary and secondary sources exist, including C. Rufus Rorem, *Blue Cross Hospital Service Plans* (Chicago: Hospital Service Plan Commission, 1944), or Odin W. Anderson, *Blue Cross Since 1929: Accountability and the Public Trust* (Cambridge, Mass.: Ballinger Press, 1975). A recent book is also recommended: Robert

Cunningham III and Robert M. Cunningham Jr., *The Blues: A History of the Blue Cross and Blue Shield System* (De Kalb: Northern Illinois University Press, 1997).

21. Starr, *Social Transformation*, 236–334; Stevens, *American Medicine and the Public Interest*, 267–89; Hendricks, *A Model for National Health Care*, 77–110.

22. Hendricks, *A Model for National Health Care*, 1.

23. For an outsider's view of organized medicine during this period and an account of his own experiment in rural medical care, see Michael Shadid's *A Doctor for the People* (New York: Vanguard Press, 1939). See also James P. Warbasse, *Cooperative Medicine: The Cooperative Organization of Health Protection* (New York: Cooperative League, 1939).

24. Secondary sources that discuss the conservative agricultural establishment's opposition to the FSA and the New Deal include Otis L. Graham, *The New Deal—The Critical Issues* (Boston: Little, Brown, 1971); Grant McConnell, *The Decline of Agrarian Democracy* (New York: Atheneum, 1969); David Conrad, *The Forgotten Farmers* (Urbana: University of Illinois Press, 1965); Larson, *Ten Years of Rural Rehabilitation*; Saloutos, *American Farmers and the New Deal*.

25. Graham, *The New Deal*, 178.

26. Hirshfield, *Lost Reform*, 84–86. Hendricks, *A Model for National Health Care*, 14–18. Starr correctly acknowledges that the FSA plans were "government-sponsored health insurance" and credits them for spurring the wider development of rural health cooperatives in the United States. Starr, *Social Transformation*, 271.

27. Davis to Alexander, 4 February 1938, Correspondence of the FSA, 1935–38, Health and Hygiene Files, Record Group 96, National Archives, Washington, D.C.

28. There is a vast scholarship that considers the role of government in medicine and public health. Among many choices, see Barbara G. Rosenkrantz, *Public Health and the State* (Cambridge, Mass.: Harvard University Press, 1972). A shorter consideration may be found in W. G. Charleton, "Government and Health Before the New Deal," *Current History* 72: 196–97, 223–24, 1977.

CHAPTER ONE: PUTTIN' OUT FOR THE FARMERS

Epigraph: Telephone interview by author, Sacramento, Calif., 24 November 1984.

1. It is a myth that the stock market crash was a singular event. The day Wall Street laid its historic "egg" was just one particularly dramatic day during which 16.5 million shares were sold and for hours not a single share was bought. Graphic descriptions of the onset and impact of the Depression are found in Robert S. McElvaine, *The Great Depression: America, 1929–1941* (New York: Times Books, 1984), 46–75, and Roger Biles, *A New Deal for the American People* (De Kalb: Northern Illinois University Press, 1991), 5–31. The sociological literature of the Great Depression alludes continually to the indelible impact this event had on the American people. Among this genre, the oral history and collected work of journalist Studs Terkel is without peer. Studs Terkel, *Hard Times: An Oral History of the Great Depression* (New York: Vintage Books, 1970).

2. Sidney Baldwin, *Poverty and Politics: The Rise and Decline of the Farm Security Administration* (Chapel Hill: University of North Carolina Press, 1968), 88. For

a discussion of the agrarian myth, see T. W. Kelsey, "The Agrarian Myth and Policy Responses to Farm Safety," *Agricultural History* 84 (1984): 1171–77, and Daniel D. Danbom, "Romantic Agrarianism in Twentieth Century America," *Agricultural History* 65 (1991): 1–12. Danbom provides a more expansive consideration of this issue in his *Born in the Country: A History of Rural Life in America* (Baltimore: Johns Hopkins University Press, 1995).

3. Walter J. Stein, *California and the Dustbowl Migration* (Westport, Conn.: Greenwood Press, 1973). Agricultural histories abound and address the profound demographic, economic, and cultural changes that are presented only briefly in the text. In addition to those cited, see also the autobiography of H. L. Mitchell, *Mean Things Happening in This Land* (Montclair, N.J.: Allanheld, Osmun, 1979). Mitchell was co-founder of the Southern Tenant Farmers Union, an interracial populist union that played an important role in agricultural politics, particularly in the South.

4. Eudora Welty, *One Time, One Place: Mississippi in the Depression* (New York: Random House, 1971), 3.

5. Paul E. Mertz, *New Deal Policy and Southern Rural Poverty* (Baton Rouge: Louisiana State University Press, 1987), 2–13.

6. Clarence Enzler, *Some Social Aspects of the Depression: 1930–1935* (Washington, D.C: Catholic University of America Press, 1939), 42.

7. American Academy of Political and Social Science, *The Medical Profession and the Public* (Philadelphia: American Academy of Political and Social Science, 1934), 21. Sydenstricker was among the first to make the astute observation that a change in socioeconomic status resulted in more severe impacts on families. Additional statistical information on the Depression is provided by two primary sources: Enzler, *Some Social Aspects of the Depression*, 17; and an exhaustive 1937 monograph by Selwyn D. Collins and Clark Tibbetts, *Research Memorandum on the Social Aspects of Health in the Depression* (New York: Social Science Research Council, 1937).

8. Arthur Raper and I. D. Reid, *Sharecroppers* (Chapel Hill: University of North Carolina Press, 1941), 37.

9. Welty, *One Time, One Place*, 3.

10. Committee on the Costs of Medical Care, *Medical Care for the American People: The Final Report of the Committee on the Costs of Medical Care*, publication no. 28 (Chicago: University of Chicago Press, 1932), cited hereafter as CCMC, *Final Report*. The CCMC envisioned an integrated system with the general practitioner at its center, serving the basic health needs of the public and making referrals to specialists or hospitals for those patients requiring greater expertise or access to high technology. Thus, concern over the decline of the generalist physician and the value of the primary care physician in a rational system of health care, a concept currently receiving a great amount of attention, first gained prominence in the deliberations of the Committee on the Costs of Medical Care.

11. CCMC, *Final Report*, 4–5.

12. Ibid., 16–22. The geographical maldistribution of medical resources began as early as the turn of the century, but it was later catalyzed, as so many social

changes were, by World War II. Recently there has been a suggestion that the out-migration of rural physicians may have plateaued and begun to reverse. *Journal of the American Medical Association* 269 (6): 744, 1993.

13. *Journal of the American Medical Association* 99 (23): 1950–52, 1932.

14. Milton I. Roemer, interview by author, Seattle, 2 October 1987.

15. Thomas Garland Moore Jr., interview by author, Sacramento, Calif., 12 February 1984.

16. Paul Taylor, *On the Ground in the Thirties* (Salt Lake City: Peregrine Smith Books, 1983), 185.

17. Carl T. Schmidt, *American Farmers in the World Crisis* (New York: Oxford University Press, 1941), 225.

18. Henry Hill Collins, *America's Own Refugees* (Princeton: Princeton University Press, 1941), 73.

19. Biles, *A New Deal*, 57–60. See also John L. Shover, *Cornbelt Rebellion: The Farmers Holiday Association* (Urbana: University of Illinois Press, 1965), and Robert E. Snyder, *Cotton Crisis* (Chapel Hill: University of North Carolina Press, 1982).

20. Taylor, *On the Ground in the Thirties*, 221–31.

21. Stein, *California and the Dustbowl Migration*, 8; *California and Western Medicine* 48 (6):107, 1938; and Henry C. Daniels, *History of the Farm Security Administration Camp Program for Migratory Agricultural Workers* (Washington, D.C.: U.S. Department of Agriculture, Farm Security Administration, Labor Division, 1944), 15.

22. Stein, *California and the Dust Bowl Migration*, 229. The movie adaptation of Steinbeck's novel *The Grapes of Wrath*, featuring a young Henry Fonda as Tom Joad, was a box office smash and so controversial that many California communities banned it. Many historians credit the novel for the decision by Congress to investigate the state of the nation's migrant work force. The investigating committee hearings may be found in U.S. Congress, House Select Committee to Investigate the Interstate Migration of Destitute Citizens, *Interstate Migration: Report of the Select Committee to Investigate the Interstate Migration of Destitute Citizens, Pursuant to H.R. 63 and H.R. 491*, 76th Cong. 2d sess.,1940.

23. John N. Webb, *The Migratory Casual Worker*, Social Science Monograph no. 7, Works Progress Administration, Division of Social Research (Washington, D.C.: GPO, 1937), 2.

24. Carey McWilliams, *Ill Fares the Land* (Boston: Little, Brown, 1942); and Carey McWilliams, *Factories in the Field: The Story of Migratory Farm Labor in California* (Boston: Little, Brown, 1939). See also Helen Johnston, *The Harvesters* (Washington, D.C.: Catholic University Press, 1968).

25. USDA, Farm Security Administration and Bureau of Agricultural Economics, *Background of the War Farm Labor Problem* (Washington, D.C.: GPO, 1942), 80–81.

26. Collins, *America's Own Refugees*, 243–51; McWilliams, *Ill Fares the Land*, 177, 286. For a consideration of the issue of venereal disease and minorities, see Allan Brandt, *No Magic Bullet: A Social History of Venereal Disease in the United States since 1880* (New York: Oxford University Press, 1987).

27. McWilliams, *Ill Fares the Land*, 173.

28. Webb, *The Migratory Casual Worker*, 2–7.

29. Ibid., 7. Also, *California and Western Medicine* 48 (6): 143, 1938.

30. Daniels, *History of the Farm Security Administration Camp Program*, 16–17. There is a large body of historiography examining societal response to disease epidemics, including the superb history of tuberculosis by Rene Dubos and Jean Dubos, *The White Plague: Tuberculosis, Man, and Society* (Boston: Little, Brown, 1952), and Charles E. Rosenberg, *The Cholera Years: The United States in 1832, 1849, and 1886* (Chicago: University of Chicago Press, 1962). Recent attempts to reconcile apparently conflicting historiographic and sociologic interpretations of the interaction between society and disease can be found in a collection of essays by Charles E. Rosenberg, *Explaining Epidemics and Other Studies in the History of Medicine* (Cambridge: Cambridge University Press, 1994), 258–318.

31. Stein, *California and the Dustbowl Migration*, 60.

32. Quoted in *California and Western Medicine* 48 (6): 143, 1938.

33. Collins and Tibbetts, *Research Memorandum on the Social Aspects of Health*, 162; Richard Hellman, "The Farmers Try Group Medicine," *Harper's Magazine*, December 1940, 5–7.

34. CCMC, *Final Report*, 22; Collins and Tibbetts, *Research Memorandum on the Social Aspects of Health*, 143–162; and Hellman, "The Farmers Try Group Medicine," 5–7.

35. Dr. William Wright, telephone interview by author, Lexington, Ky., 24 November 1979.

36. B. K. Harris, "Plow Beams for Pills," in *Federal Writers' Project: These Are Our Lives* (Chapel Hill: University of North Carolina Press, 1939), 269.

37. Thomas Garland Moore Jr., interview by author, Sacramento, Calif., 24 November 1979.

38. Early writers on the historiography of the New Deal were inclined to see FDR and New Deal liberalism as the saviors of American society. Typical of this view is the well-known three-volume biography of Roosevelt by Arthur M. Schlesinger Jr., *The Age of Roosevelt: The Crisis of the Old Order, 1919–1933* (Boston: Houghton Mifflin, 1957); *The Age of Roosevelt: The Coming of the New Deal* (Boston: Houghton Mifflin, 1959); and *The Age of Roosevelt: The Politics of Upheaval* (Boston: Houghton Mifflin, 1960). In the 1960s, this uncritical view was supplanted by the work of various New Left historians who critiqued the New Dealers for their failure to fully resolve the social and economic problems that led to the crisis in the first place. Typical of this body of work are Paul Conkin's *The New Deal* (Arlington Heights, Ill.: AHM Publishing, 1967), and Otis L. Graham, *The New Deal: The Critical Issues* (Boston: Little, Brown, 1971). Recent books include Kenneth S. Davis, *FDR: The New Deal Years, 1933–1937* (New York: Random House, 1986). Two previously cited books are also recommended: McElvaine, *The Great Depression*, and Biles, *A New Deal*.

39. Biles, *A New Deal*, 96–116. The Brain Trust provided the substance for Roosevelt's campaign themes and thereby created the substrate on which the New Deal was built. Tugwell, one of the most visionary and outspoken of Roosevelt's

braintrusters, has attracted much historical attention. One of several good Tugwell biographies is Bernard Sternsher, *Rexford Tugwell and the New Deal* (New Brunswick, N.J.: Rutgers University Press, 1964). Joining Tugwell were two colleagues from Columbia, Raymond Moley, a political scientist, and Adolf H. Berle Jr., a liberal academic who supported business-government cooperation in economic planning. Tugwell brought with him to the USDA a cadre of liberal urban lawyers, a fact that rankled the traditional agricultural establishment whose roots were in the soil, not the ivory tower. A fine biography of Wallace has recently been published: Graham J. White and John Maze, *Henry A. Wallace: His Search for a New World Order* (Chapel Hill: University of North Carolina Press, 1994).

40. Daniel S. Hirshfield, *The Lost Reform: The Campaign for Compulsory National Health Insurance from 1932 to 1943* (Cambridge, Mass.: Harvard University Press, 1970), 61.

41. H. Hopkins, "Unemployment Relief and the Public Welfare Policy," *Proceedings of the American Academy of Political and Social Science* 15 (January 1934): 81. Hopkins' multiple and evolving roles in New Deal domestic policy are covered in Biles, *A New Deal*, 96–116. A superb biography of Hopkins was written by George McJimsey, *Harry Hopkins: Ally of the Poor and Defender of Democracy* (Cambridge, Mass.: Harvard University Press, 1987).

42. Doris Carothers, *Chronology of the Federal Emergency Relief Administration* (New York: Da Capo Press, 1937, reprint ed., 1971). See also Theodore E. Whiting, *Final Statistical Report of the Federal Emergency Relief Administration* (Washington, D.C.: GPO, 1942) and George St. John Perrott, Edgar Sydenstricker, and Selwyn D. Collins, "Medical Care during the Depression," *Milbank Memorial Fund Quarterly* 12 (2): 99–114, 1934.

43. These data were included in the report issued in 1938 by the Technical Committee on Medical Care. See Interdepartmental Committee to Coordinate Health and Welfare Activities, *A National Health Program: Report of the Technical Committee on Medical Care, Proceedings of the National Health Conference, July 18,19,20, 1938* (Washington, D.C.: GPO, 1938), and U.S. Public Health Service, *The Nation's Health, Report of the Technical Committee on Medical Care* (Washington, D.C.: GPO, 1938), 5–7.

44. Carothers, *Chronology of the Federal Emergency Relief Administration*, 5–18.

45. American Academy of Political and Social Science, *Medical Care for Americans* (Philadelphia: American Academy of Political and Social Science, 1951), 86.

46. *Journal of the Missouri Medical Association* 30 (12): 496, 1933.

47. Ibid., 30 (7): 295–98, 1933 and 30 (12): 496, 1933.

48. Thomas Garland Moore Jr., interview by author, Sacramento, Calif., 12 February, 1984. The literature on the New Deal's and the FSA's record on race issues is mixed. No doubt abuses by some FSA officials occurred at the local level. The Southern Tenant Farmers Union, one of the few explicitly integrated unions of that era, complained bitterly about New Deal policies that it thought systematically excluded the poorest, least-productive farm families. Black farmers would be overrepresented in this group. On the whole, however, the FSA's record is con-

siderably better than most New Deal agencies. Saloutos, *The American Farmer and the New Deal*, 185–87; Biles, *A New Deal*, 172–92.

49. Mertz, *New Deal Policy and Southern Rural Poverty*, 58. When secretary of agriculture Henry Wallace was trying to decide whether to bow to pressure from organized agricultural groups to scale back or eliminate the FSA's resettlement and tenant purchase programs, he went on a car tour of the South's "Tobacco Roads," and returned to Washington determined to continue the controversial efforts. On Eleanor Roosevelt and race, see Kearns Goodwin, *No Ordinary Time*, 162–65.

50. Cited in Mertz, *New Deal Policy and Southern Rural Poverty*, 2.

51. The AAA is covered in great detail by Kirby, *Rural Worlds Lost*; Saloutos, *American Farmer and the New Deal*; and Grant McConnell, *The Decline of Agrarian Democracy* (New York: Atheneum, 1969).

52. Lester V. Chandler, *America's Greatest Depression, 1929–1941* (New York: Harper and Row, 1984), 209; Kirby, *Rural Worlds Lost*, 50; Mertz, *New Deal Policy and Southern Rural Poverty*, 57.

53. This consolidation brought into the FERA two additional programs that touched on rural rehabilitation issues: the National Industrial Recovery Administration's Division of Subsistence Homesteads (under Harold Ickes' control as secretary of the interior), and the Farm Credit Administration of the Department of Agriculture. Although the new programs had national intent, southern farmers received 93 percent of the funds disbursed by the Division of Rural Rehabilitation and Stranded Populations. These events are summarized in Conkin, *The New Deal*, 57–97; McElvaine, *The Great Depression*, 126–28; and Mertz, *New Deal Policy and Southern Rural Poverty*, 80–81. See also Berta Asch and A. R. Mangus, *Farmers on Relief and Rehabilitation* (Washington, D.C.: GPO, 1937), 17–21.

54. Theodore Rosengarten, *All God's Dangers: The Life of Nate Shaw* (New York: Alfred A. Knopf, 1975), 287 and 291.

55. For a discussion of the creation of the RA, see Baldwin, *Poverty and Politics*, 90–94; Davis, *FDR: The New Deal Years*, 471–76; and Conkin, *The New Deal*, 57–58.

56. Conkin, *The New Deal*, 57–58. Tugwell appointed Alexander to head what became one of the agency's most important programs, the Division of Rural Rehabilitation. Alexander later replaced Tugwell as the RA Administrator in January 1937 just before the RA became the FSA. Two other divisions whose programs were innovative and controversial—and by all accounts, woefully underfunded—were the Tenant Purchase Division and the Resettlement Division. Davis, *FDR: The New Deal Years*, 471–90.

57. While these communities received federal subsidies, they were autonomously governed by elected community councils. FSA managers assigned to the communities exercised considerable control over which families were eligible for residence and other community activities and policies. By World War II, nearly 11,000 homesteads had been built under the subsistence homestead program. There were a half-dozen homestead projects funded by the FSA, including Tygart Valley and Arthurdale in West Virginia, the Westmoreland project in Pennsylvania, and Aberdeen Gardens, an all-black community near Newport News, Virginia. A

total of three greenbelt communities were built by the agency in Greenbelt, Maryland; Greendale, Wisconsin; and Greenville, Ohio. The greenbelt programs were federally administered communities until 1955, when they were sold to private developers. Greenbelt communities, subsistence homesteads, and resettlement projects are discussed in Baldwin, *Poverty and Politics*, 105–6, 187–89, 214–17; T. H. Watkins, *The Great Depression: America in the 1930s* (Boston: Little, Brown, 1993), 261–62; and White and Maze, *Henry A. Wallace*, 75–80. A more thorough history of these model communities is found in Paul Conkin's *Tomorrow a New World: The New Deal Community Program* (Ithaca, N.Y.: Cornell University Press, 1959). For brief comments on the resettlement communities' medical care programs, see Frederick D. Mott and Milton I. Roemer, *Rural Health and Medical Care* (New York: McGraw-Hill, 1948), 401.

58. Thomas Garland Moore Jr., interview by author, Sacramento, Calif., 12 February 1984.

59. Baldwin, *Poverty and Politics*, 284. The "purge" of liberals in the Department of Agriculture in January 1935 was in part due to the battle over maintaining tenants under AAA policies and in part due to these individuals' fervent support of the more utopian ideas of Tugwell and his disciples in the USDA. See Davis, *FDR: The New Deal Years*, 478–81.

60. David Conrad, *The Forgotten Farmers* (Urbana: University of Illinois Press, 1965), 105–19.

61. Pynchon to West, 3 May 1935, Correspondence of James Heizer, Records of the FSA, Region IV, Office of the Director, Medical Services Files 936, FSA Correspondence 1935–42, Record Group 96, National Archives (hereafter cited as Heizer Correspondence).

62. West to Pynchon, 18 May 1935, Heizer Correspondence. For a discussion of organized medicine's opposition to contract practice, see George Rosen, "Contract or Lodge Practice and its Influence on Medical Attitudes to Health Insurance," *American Journal of Public Health* 67 (4): 374–78, 1977.

63. *Journal of the American Medical Association*, organizational section 108 (14): 111B–112B, 1936.

64. *New England Journal of Medicine* 209 (12): 586–91, 1933.

65. I. S. Falk was a CCMC veteran and a fixture in federal health policy circles throughout the Roosevelt presidencies, remaining an influential voice for decades. Sydenstricker was a statistician with the USPHS who later moved on to become medical director of the Milbank Memorial Fund. Sydenstricker co-authored one of the CCMC's minority reports in which he recommended compulsory insurance. See I. S. Falk, "Medical Care in the USA—1932–1972. Problems, Proposals, and Programs from the Committee on the Cost of Medical Care to the Committee for National Health Insurance," *Milbank Memorial Fund Quarterly* 51 (1):1-31, 1973.

66. Biles, *A New Deal*, 11; *Journal of the Iowa State Medical Society* 27 (9): 496–98, 1937; *Journal Lancet* 58 (9): 395–97, 1938.

67. U.S. Resettlement Administration, *Annual Report Office of the Medical Director, Resettlement Administration for Fiscal Year Ending June 30, 1937* (Wash-

ington, D.C.: GPO, 1937), file 31-19, FSA/ RA Office of the Medical Director, Reports 1936–1937, Frederick Dodge Mott Papers, Canadian National Archives, Ottawa, Canada.

68. Ibid., 5–13.

69. Ibid.

70. Ibid., 11–13. Doctors and hospitals received 51 and 43 percent of all funds, respectively. Events in North and South Dakota are covered in the medical journal shared by the two state medical societies, the *Journal Lancet.*

71. *Journal of the American Medical Association*, organizational section, 109 (10): 28B, 1937.

72. Ibid., organizational section 108 (18): 1518, 1937.

73. Ibid., organizational section 109 (10): 28B, 1937.

74. U.S. Resettlement Administration, *Annual Report Office of the Medical Director*, 13.

75. *Journal of the American Medical Association*, organizational section 108 (18): 1518, 1937.

76. Ibid.

77. *Journal of the Iowa State Medical Association* 27 (9): 493–98, 1937; see also *Journal of the American Medical Association*, organizational section 108 (14): 107b–109b, 1937.

78. *Journal of the Iowa State Medical Association* 27 (9): 493–98, 1937.

79. Ibid. Initially families were given promissory notes, which essentially transferred risk of collection from clients to the FSA. This plan was suspended by the FSA because of escalating costs. It then introduced a plan to pay $1 per month with more prorating. This led to abrogation of the agreements by both state medical societies. In January 1938, the FSA tried again, introducing a $24 per year per capita fund from which discounted fees were paid. The next year this sum was increased to $33 per year and because of improved farm prices it was limited to the ten most severely affected counties. USDA, Farm Security Administration, *Office of the Chief Medical Officer Progress Report for 1939* (Washington, D.C.: GPO, 1939), 27–30, file 31-17, and USDA, Farm Security Administration, *Annual Report of the Chief Medical Officer, Fiscal Year Ending June 30, 1939* (Washington, D.C.: GPO, 1939), file 31-13. Both located in Frederick Dodge Mott Papers, Canadian National Archives, Ottawa.

80. *Journal Lancet* 59 (7): 330–34, 1939.

81. Ibid., 58 (7): 318, 344–45, 356, 1938 and *Journal of the American Medical Association* 110 (8): 113b–114b, 1938.

82. U.S. Resettlement Administration, *Interim Report of the Resettlement Administration* (Washington, D.C.: GPO, 1936), 93–94.

CHAPTER TWO: A MATTER OF GOOD BUSINESS

Epigraph: Telephone interview by author, Sacramento, Calif., 24 November 1979.

1. Theodore Saloutos, *The American Farmer and the New Deal* (Ames: Iowa State University Press, 1982), 162–63; Roger Biles, *A New Deal for the American People* (De Kalb: Northern Illinois University Press, 1991), 75; Daniel D. Danbom,

Born in the Country: A History of Rural Life in America (Baltimore: Johns Hopkins University Press, 1995), 219–20; John T. Kirby, *Rural Worlds Lost: The American South, 1920–1960* (Baton Rouge: Louisiana State University Press, 1987), 56–67.

2. Sidney Baldwin, *Poverty and Politics: The Rise and Decline of the Farm Security Administration* (Chapel Hill: University of North Carolina Press, 1968), 95–96.

3. The position of chief medical officer was briefly held by Dr. Robert Oleson, who was replaced by Williams in January 1936. Although the FSA typically chose regional directors who were acceptable to conservative farm interests, all middle managerial positions were filled by the Washington office. The same was true for the medical care programs. This gave the agency's Washington staff considerable influence over who would put forward the agency's programs in the field. U.S. Resettlement Administration, *Annual Report—Office of the Medical Director, Resettlement Administration for Fiscal Year Ending June 30, 1937* (Washington, D.C.: GPO, 1937). Also, Mott to Health Services Staff, 13 July 1944, Frederick Dodge Mott Papers, Canadian National Archives, Ottawa, Canada.

4. Alexander to Mitchell, 17 August 1937, Correspondence of James Heizer, Records of the FSA, Region IV, Office of the Director, Health and Rehabilitation files, FSA Correspondence 1935–1942, Record Group 96, National Archives (hereafter cited as Heizer Correspondence with files noted when available).

5. Browning to Swift, 11 September 1937, Health and Rehabilitation files, Heizer Correspondence.

6. Lamb to Slack, 6 October 1937, Health and Rehabilitation files, Heizer Correspondence.

7. USDA, Farm Security Administration, *Group Medical Care for Farmers*, Publication no. 75 (Washington, D.C.: GPO, 1941), 3–4.

8. Mitchell to Alexander, 16 October 1937, Health and Rehabilitation files, Heizer Correspondence.

9. Browning to Swift, 11 September 1937, Health and Rehabilitation files, Heizer Correspondence.

10. Thomas Garland Moore Jr., interview by author, Sacramento, Calif., 12 February 1984. The same point was made in an earlier interview with Moore, telephone interview by author, Sacramento, Calif., 24 November 1979.

11. Ware to Slack, 15 October 1937, Health and Rehabilitation files, Heizer Correspondence.

12. Ibid. See also Hindle to Slack, 5 October 1937 and Hamilton to Swift, 6 September 1937, Health and Rehabilitation files, Heizer Correspondence.

13. USDA, Farm Security Administration, *Medical Care for Farm Security Administration Borrowers, June 15, 1939* (Washington, D.C.: GPO, 1939), 2.

14. R. C. Williams, "Digest of a Speech Delivered to the American Public Health Association," Paper presented at the Annual Meeting of the APHA, New York City, 1 November 1939, file 31-19, Frederick Dodge Mott Papers, Canadian National Archives, Ottawa.

15. *Raleigh News and Observer*, 14 July 1937, Health Miscellaneous files, Heiz-

er Correspondence. According to this report, six state physicians, including the state health director, had supported the Committee of 400, whose manifesto proclaimed that the health of the people was a direct concern of the government. This prompted a special meeting of the state medical society to discuss what action it would take against the "rebels."

16. Beginning in 1937, at least sixteen letters passed back and forth between Washington and North Carolina, as the FSA tried to put into place a statewide memorandum of understanding. Williams to Long, 27 December 1938, Health and Hygiene files, North Carolina, 1935–June 1939, Region IV General Correspondence, 1935–1940, Farm Security Administration, Record Group 96, National Archives (hereafter cited as North Carolina Health and Hygiene files).

17. Houser to Williams, 17 November 1937, files of the Medical Service, North Carolina, 1935–June 1939, FSA Region IV Correspondence, Record Group 96, National Archives. For a discussion of states' rights in the South, see John Gaventa, *Power and Powerlessness: Quiescence and Rebellion in an Appalachian Valley* (Urbana: University of Illinois Press, 1980).

18. Williams to Johnson, 1 December 1937, Health and Rehabilitation files, Heizer Correspondence.

19. Williams to Long, 27 December 1938, North Carolina Health and Hygiene files.

20. *Memorandum of Understanding, Robeson County, North Carolina,* appendix F, in Michael R. Grey, "Primary Care in the Great Depression" (B.A. thesis, Harvard College, 1980).

21. Hill to O'Daniel, 28 October 1942, Cass County Rural Health Association, Records of the Cooperative Associations, Texas, Correspondence of the Farmers Home Administration, 1935–1954, Record Group 96, National Archives.

22. Williams to Gordon, 5 January 1940, North Carolina Health and Hygiene files. There were tie-ins to other joint efforts, such as the inclusion of FSA clients in a joint state medical society and state hospital association group hospitalization program in North Carolina, called the Hospital Savings Association. Mott to Williams, 23 June 1937, Heizer Correspondence.

23. Tugwell was visionary in this respect, and he started the Information Division program anticipating the controversy his resettlement efforts would elicit. There is an extensive literature about the photographs taken by the Historical Division of the FSA. The group was directed by Roy Stryker, and it was Stryker who brought what is probably the finest group of American photographers ever assembled in one program, including such giants of American photography as Dorothea Lange, John Vachon, Walker Evans, Arthur Rothstein, Ben Shahn, Margaret Bourke-White, Russell Lee, Gordon Parks, Jack Delano, Carl Mydans, Marion Post Wolcott, and John Collier. Edward Steichen, who had once been the director of photography at the Museum of Modern Art in New York City, was of the opinion that the work of the FSA photographers produced "the most remarkable human documents that were ever rendered in pictures." Baldwin, *Poverty and Politics,* 117–19.

24. Slack to Stewart, 10 October 1939, North Carolina Health and Hygiene files.

25. Hoover to Wood, 10 June 1939, North Carolina Health and Hygiene files.

26. The FSA medical care programs are detailed in Frederick Dodge Mott and Milton I. Roemer, *Rural Health and Medical Care* (New York: McGraw-Hill, 1948), 312–432; USDA, Farm Security Administration, *Group Medical Care for Farmers*; and Swift to Slack, 8 December 1937, Heizer Correspondence.

27. Mott and Roemer, *Rural Health and Medical Care,* 400.

28. USDA, Farm Security Administration, *Group Medical Care for Farmers,* 5.

29. Alec Spencer, telephone interview by author, Lexington, Ky., 24 November 1979.

30. Elizabeth Fee and Dorothy Porter, "Public Health, Preventive Medicine, and Professionalization," in Andrew Wear (ed.), *Medicine in Society: Historical Essays* (Cambridge: Cambridge University Press, 1991), 249–75.

31. This development indicates that the FSA medical care program included features suggestive of peer review organizations and independent practice associations that would become common in the last quarter of the century, a point noted in passing by Paul Starr, *The Social Transformation of American Medicine* (New York: Basic Books, 1976), 325.

32. USDA, Farm Security Administration, *Annual Report for the Fiscal Year, July 1, 1939 to June 30, 1940* (Washington, D.C.: GPO, 1940), 23–24. In North Carolina, FSA Director Vance Swift adopted a similar policy in 1937, causing an upsurge in membership there. This was also seen in regions III and V. Maguire to All Regional Directors, 10 August 1936, Health and Rehabilitation files, Heizer Correspondence.

33. Mott and Roemer, *Rural Health and Medical Care,* 389–432.

34. John Newdorp, a Public Health Service physician who was FSA regional medical director in Atlanta, later recalled one early experience that caused him considerable personal anguish: "When I first came there was a retired physician who was a surgeon and the situation was not one where people were getting very good surgical care, at least. I had to replace him, it was one of the early things that I had to do. Not a very pleasant job. . . all kinds of repercussions all the way up to Washington." John Newdorp, interview by author, Alexandria, Va., 3 February 1984.

35. USDA, Farm Security Administration, *Group Health Services Manual,* Region IV, February 1941, 3–16, file 31-4, Frederick Dodge Mott Papers, Canadian National Archives, Ottawa.

36. Ibid., 12–16 (emphasis in original).

37. Maguire to All Regional Directors, 4 May 1938, Health and Rehabilitation files, Heizer Correspondence.

38. Grey, "Primary Care in the Great Depression," Health Plan Participation Plan, appendix H. This document was originally copied from the personal papers lent to me by Dr. Mott for my undergraduate thesis research. Later I was unable to locate it among Mott's papers in the Canadian National Archives.

39. Thomas Garland Moore Jr., telephone interview by author, Sacramento, Calif., 11 November 1979.

40. Mott and Roemer, *Rural Health and Medical Care,* 413.

41. T. H. Watkins, *The Great Depression: America in the 1930s* (Boston: Little, Brown, 1993), 261–62. Between 1935 and 1942, approximately forty resettlement communities, housing 4,411 families, had been built. A total of 11,000 homesteads were built under the subsistence homestead program, and three of four planned suburban greenbelt towns were built. For additional discussion of resettlement, greenbelt, and homestead communities of the RA and the FSA, see Baldwin, *Poverty and Politics,* 111–13, 214–17.

42. Baldwin, *Poverty and Politics,* 284.

43. USDA, Farm Security Administration, *Summary of Annual Report,* June 1938 (Washington, D.C.: GPO, 1938).

44. Mott and Roemer, *Rural Health and Medical Care,* 401–2. See also USDA, Farm Security Administration, *Annual Report of the Chief Medical Officer for the Fiscal Year Ending June 30, 1939* (Washington, D.C.: GPO, 1939), 24–26. Over time there was a tendency to merge these resettlement project plans with the medical care unit for standard rehabilitation borrowers. The FSA encouraged this development, since it increased administrative efficiency and helped to decrease confusion by local physicians.

45. USDA, Farm Security Administration, *Annual Report of the Chief Medical Officer for the Fiscal Year Ending June 1939;* Ralph C. Williams, *Summary of the Annual Report Office of the Medical Director, Farm Security Administration, July 1, 1938–June 30, 1938* (Washington, D.C.: GPO, 1938); and Mott and Roemer, *Rural Health and Medical Care,* 400–421.

46. USDA, Farm Security Administration, *Group Medical Care for Farmers,* 11; and Olaf F. Larson, *Ten Years of Rural Rehabilitation in the United States* (Bombay, India: Association of Advertisers and Printers, 1950), 90–91. In Missouri, the number of calls ranged from 99 to 165 per 1,000 members and collections averaged just short of 70 percent. *Journal of the Missouri Medical Association* 40 (8): 259–63, 1943, and *Journal of the Missouri Medical Association* 41 (7): 139, 1944; Richard Hellman, "The Farmers Try Group Medicine," *Harper's Magazine,* December 1940, 7; and USDA, Farm Security Administration, *Group Health Services Manual Region IV,* February 1941, 12, file 31-4, Mott Papers.

47. *Journal of the Medical Association for the State of Alabama* 11 (7): 242–45, 1942, and *Journal of the Missouri Medical Association* 40 (7): 224, 1943.

48. Lorin Kerr, interview by author, Washington, D.C., 11 November 1983.

49. *Texas State Medical Journal* 38 (2): 20-22, 1942, and *Texas State Medical Journal* 37 (2): 106–9, 1941.

50. *Journal of the American Medical Association,* organizational section 114 (2): 164, 1940.

51. Cited in *California and Western Medicine* 47 (6): 432–33, 1937.

52. Many FSA Health Services personnel, including such senior staff members as Ralph Williams and his assistant Fred Mott, were commissioned officers in the corps. The FSA also relied on sanitary engineers and nurses drawn from the U.S.

Public Health Service in its national, regional, and district field offices. A summary of the U.S. Public Health Service during the Great Depression and World War II is found in Fitzhugh Mullan, *Plagues and Peoples: The Story of the United States Public Health Service* (New York: Basic Books, 1989), 82–126.

53. "A National Health Program: Report of the Technical Committee on Medical Care, Interdepartmental Committee to Coordinate Health and Welfare Activities, Proceedings of the National Health Conference, July 18, 19, 20, 1938" (Washington, D.C.: GPO, 1938), 11. For a general discussion of these events, see Starr, *Social Transformation,* 273–77; Rosemary Stevens, *American Medicine and the Public Interest* (New York: Basic Books, 1987), 175–97, 267–89; or the more comprehensive book by Daniel Hirshfield, *The Lost Reform: The Campaign for Compulsory National Health Insurance from 1932 to 1943* (Cambridge, Mass.: Harvard University Press, 1971). A contemporary view of the turmoil within the medical profession around national health insurance can be found in James Rorty, *American Medicine Mobilizes* (New York: W. W. Norton, 1939), although blow-by-blow accounts are readily available in virtually any medical journal, especially the *Journal of the American Medical Association.*

54. *Journal of the Medical Association of the State of Alabama* 8 (12): 428–29, 1939.

55. *Journal of the American Medical Association,* organizational section 120 (16): 1317, 1942.

56. *Journal of the Medical Association of the State of Alabama* 8 (12): 432–33, 1939.

57. *Journal of the Indiana State Medical Association* 30 (1): 32, 1937. The FSA made little headway in setting up medical care cooperatives in Delaware and Maryland, although the migrant health program had a substantial presence in both states. There were a few notable exceptions along the East Coast and into New England. For example, the Vermont Medical Association agreed to support a statewide medical care plan in 1938 mainly because there were so few rural rehabilitation borrowers in the state and they were widely dispersed. Connecticut had a very small FSA clientele, although the agency provided medical care to migrant workers scattered throughout the state during the war years. In New Jersey, the state association approved a statewide plan in 1938 and participated in both the medical cooperative and migrant health programs for over six years. USDA, Farm Security Administration, *Annual Report of the Chief Medical Officer* (1939, 1940).

58. *Pennsylvania Medical Journal* 43 (8): 1185, 1940.

59. *Journal of the American Medical Association* 110 (17): 196b, 1938; *California and Western Medicine* 56 (6): 373, 1942.

60. Mott to Williams, 30 June 1938, Heizer Correspondence.

61. Smith to Alexander, 15 April 1937, FSA Correspondence 1935–1938, Sterling Library, Yale University, New Haven, Conn.

62. *Journal of the American Medical Association* 120 (16): 1315–24, 1942.

63. Harold Mayers, interview by author, Washington, D.C., 3 February 1984.

64. Pickle to Goff, 21 April 1939, North Carolina Health and Hygiene files.

65. USDA, Farm Security Administration, *Annual Report of the Chief Medical Officer* (1940), 10–12; USDA, Farm Security Administration, *Annual Report of the Chief Medical Officer for the Fiscal Year July 1, 1940–June 30, 1941* (Washington, D.C.: GPO, 1941), 19–40.

66. Meriwhether to Gordon, 3 November 1939, North Carolina Health and Hygiene files.

67. Goff to Swift, 21 July 1939, North Carolina Health and Hygiene files.

68. Meriwhether to Gordon, 3 November 1939, North Carolina Health and Hygiene files.

69. *Journal of the American Medical Association* 115 (3): 224, 1940.

70. *Journal of the Medical Association of the State of Alabama* 11 (7): 242–45, 1942.

71. USDA, Farm Security Administration, *Annual Report of the Chief Medical Officer* (1940), 21.

72. USDA, Farm Security Administration, *Annual Report of the Chief Medical Officer* (1941), 38–40.

73. *Journal of the Medical Association of the State of Alabama* 11 (7): 242–45, 1942.

74. *Journal of the American Medical Association* 114 (2): 159–67, 1940; *Journal of the Arkansas Medical Society* 37 (7): 149, 1940; *Journal of the Iowa State Medical Society* 27 (7): 324–51, 1937.

75. *Journal of the American Medical Association* 114 (2): 159–167, 1940.

76. *Texas State Medical Journal* 39 (2): 88, 1943 and 40 (1): 83–85, 1944. One society replied in the survey that its members felt coercion from the FSA, writing to the state society that a "threat has been made by Dr. Mott of Washington . . . to the effect that unless the doctors continued their work at reduced fees, the government would draft doctors and put them in these camps. . ." Obviously in this case the comment referred to the migrant camp health programs, and not the cooperatives, although both operated in Texas. *Texas State Medical Journal* 39 (2): 88, 1943.

77. *Journal of the American Medical Association* 108 (14): 107b–109b, 1937.

78. Slack to Rauscher, 15 July 1942, Records of the Cooperative Associations, Tennessee, Franklin County Medical Care Association, Record Group 96, National Archives. Also, Bates to Johnston, 24 March 1938, Heizer Correspondence.

79. *Journal of the American Medical Association* 111 (25): 2307–8, 1938.

80. David J. Rothman, "A Century of Failure: Health Care Reform in America," *Journal of Health Politics, Policy, and Law* 18 (2): 271–85, 1993; Roy Lubove, "The New Deal and National Health," *Current History* 78 (427): 198–226, 1977; Hirshfield, *Lost Reform*, 152–53; and Starr, *Social Transformation*, 275–89.

81. *California and Western Medicine* 50 (1): 20–21, 1939.

82. *Journal of the American Medical Association* 114 (2): 159–67, 1940.

83. Ibid.

84. Saloutos, *The American Farmer and the New Deal*, 267.

85. There are various estimates of this repayment figure, but they range narrowly between 80 percent and the 90 percent figures cited in the text. In some re-

gions, the agency tended to bypass the least productive farmers in an effort to boost repayment rates, a fault that has nettled many historians since. Baldwin, *Poverty and Politics,* 138; Biles, *A New Deal,* 75–77. See also Theodore Saloutos, "New Deal Agricultural Policy: An Examination," *Journal of American History* 61 (9): 416, 1974.

86. Elizabeth Etheridge, *The Butterfly Caste: A Social History of Pellagra in the South* (Westport, Conn.: Greenwood Press, 1972), 205. There are a number of sources for statistical data on rural health in this period. The most detailed include the series *Public Health Reports,* often authored by statisticians from the U.S. Public Health Service Statistical division, who were allies of the FSA, such as George St. John Perrott, Jesse Yaukey, and Joseph Mountin. See for example, *Public Health Reports* 59 (36): 1163–84, 1944; 60 (16): 429–46, 1945; and 60 (25): 625–710, 1945.

87. Larson, *Ten Years of Rural Rehabilitation,* 89–90. See also C. E. Lively and F. Lionberger, *Physical Status and Health of Farm Tenants and Farm Laborers in Southeast Missouri* (Columbia: University of Missouri Press, July 1942). This report was a joint effort of the University of Missouri's Department of Rural Sociology, the Missouri Agricultural Extension Service, the FSA (Region III), and several private agencies, including the Missouri State Medical Association. Additional primary sources on the FSA's nutrition and sanitation efforts include the annual reports of the chief medical officer previously cited.

88. Etheridge, *Butterfly Caste,* 205.

89. Cited in Larson, *Ten Years of Rural Rehabilitation,* 90.

90. USDA, Farm Security Administration, *Group Medical Care for Farmers,* 11.

91. USDA, Farm Security Administration, *Annual Report of the Chief Medical Officer* (1939), 27.

92. Richard Couto, *Ain't Gonna Let Nobody Turn Me Around: The Pursuit of Racial Justice in the Rural South* (Philadelphia: Temple University Press, 1991), 266.

CHAPTER THREE: A LONG AND FAR-REACHING PROGRAM

Epigraph: Woody Guthrie, "Pastures of Plenty," TRO Copyright, Ludlow Music, Inc., with permission.

1. U.S. Department of Labor, Retraining and Reemployment Administration, *Federal Interagency Committee on Migrant Labor Report and Recommendations: Migrant Labor . . . A Human Problem* (Washington, D.C.: GPO, March 1947), 15; U.S. Department of Labor, *History of the F.S.A. Camp Program for Migrant Agricultural Workers* (Washington, D.C.: GPO, 1944).

2. Lorin Kerr, interview by author, Washington, D.C., 11 November 1983.

3. Walter J. Stein, *California and the Dustbowl Migration* (Westport, Conn.: Greenwood Press, 1973), 156–57.

4. John Steinbeck, *The Grapes of Wrath* (New York: Vintage Books, 1938).

5. Frederick D. Mott and Milton I. Roemer, *Rural Health and Medical Care* (New York: McGraw-Hill, 1948), 424–25; Wayne Rasmussen, *The History of the Farm Labor Programs of the U.S. Department of Agriculture, 1935–1947* (Wash-

ington, D.C.: GPO, 1949); Bureau of Agricultural Economics and Farm Security Administration, *Background of the War Farm Labor Problem* (Washington, D.C.: May 1942); and U.S. Department of Labor, *History of the F.S.A. Camp Program for Migrant Agricultural Workers* (Washington, D.C.: GPO, 1944), 47–55. A fine regional history of migrant labor in the Connecticut River valley is Fay Clarke Johnson's *Soldiers of the Soil* (New York: Vantage Press, 1995).

6. Henry C. Daniels, interview by author, Takoma Park, Md., 21 December 1984.

7. Thomas Garland Moore Jr., interview by author, Sacramento, Calif., 12 February 1984.

8. Lorin Kerr, interview by author, Washington, D.C., 11 November 1983; Henry C. Daniels, interview by author, Takoma Park, Md., 21 December 1984; Thomas Garland Moore Jr., interview by author, Sacramento, Calif., 12 February 1984.

9. Carey McWilliams, *Ill Fares the Land* (Boston: Little, Brown, 1942), 48.

10. The federal government's involvement in migrant welfare issues is the subject of a number of primary and secondary sources. The essays of journalist Paul Taylor (the husband of photographer Dorothea Lange) are particularly graphic and spare. Paul Taylor, *On the Ground in the Thirties* (Salt Lake City: Peregrine Smith Books, 1983), especially 213–20.

11. The 1930s were a true watershed in the history of hospital care in the United States. Especially notable was the emergence of hospital insurance plans, such as Blue Cross, that put the cost of hospitalization within the reach of ordinary citizens. According to Rosenberg, by the 1920s the modern hospital had already been transformed into the central health care entity that we know today. While this was true enough in the nation's major urban centers, in rural America hospitals remained few and far between and played a much more limited role in day-to-day medical practice. Not until the passage of the Hill-Burton Act in 1946, which provided a massive infusion of federal funding and caused an explosion in hospital construction, would the hospital become a fixture in rural communities. Two comprehensive histories of hospitals in America are Charles E. Rosenberg, *The Care of Strangers* (New York: Basic Books, 1987), and Rosemary Stevens, *In Sickness and in Wealth* (New York: Basic Books, 1994).

12. The California state health department had been sending nurses and physicians on immunization campaigns for several years before the FSA began the AWHMA in that state. Upward of 60,000 families benefited from these efforts, which were designed to stem the risk of communicable disease in communities where there was an influx of migrants. Other state agencies, such as the California Department of Housing and Immigration, could not keep up with the sheer numbers in order to adequately inspect grower camps and the myriad private squatter camps that where everywhere in evidence. See Bureau of Agricultural Economics and Farm Security Administration, *Background of the War Farm Labor Problem*, 110–15.

13. Mills to Garst, 3 February 1938 in "A Health Program for Migratory Agri-

cultural Workers in California," file 243, Records of the FSA, Bancroft Library, University of California, Berkeley.

14. *California and Western Medicine* 48 (6): 397, 460–61, 1937. The same approach was used in other AWHAs. Daniels recalled that the board of directors for the Texas Farm Laborers Health Association had "people who were interested in migrant programs, public-spirited citizens sometimes, local people and that kind of thing. The minority of them would be representatives of the Farm Security Administration or the government." Near the end of the war, Daniels was national administrator for the AWHAs and was the liaison between the associations and the FSA. Henry C. Daniels, interview by author, Takoma Park, Md., 21 December 1984.

15. Mitchell to Duffy, 11 September 1940, FSA General Correspondence, 1935–1942, Record Group 96, National Archives and Records Administration Region XI, Seattle, Wash.

16. The Atlantic Seaboard AWHA initially covered Florida as well as the states of North Carolina, Virginia, Maryland, Delaware, Pennsylvania, New Jersey, New York, and Connecticut. Florida later developed its own AWHA, the Migrant Labor Health Association. Records for the War Food Administration are classified as Record Group 224 and are cited as such.

17. U.S. Department of Agriculture Production and Marketing Administration, Labor Branch, "Agricultural Workers Health Associations Report of Activities, January–December, 1946" (Washington, D.C.: GPO, 1947).

18. Migratory Labor Health Association, *Nurses Manual* (Atlanta, 1944), 1–3. Located in personal papers of Henry C. Daniels, loaned to author.

19. "Clinics on Wheels," *New York Herald Tribune,* 23 October 1941.

20. Bureau of Agricultural Economics and Farm Security Administration, *Background of the War Farm Labor Program,* 112–16.

21. Duffy to Foley, 19 September 1940, FSA General Correspondence, 1935–1942, Record Group 96, National Archives and Records Administration, Federal Records Center, Region XI, Seattle, Wash.

22. War Food Administration, Office of Labor, *Agricultural Workers Health Association Clinic Procedure Manual, Region XI* (War Food Administration, Office of Labor, April 1942), 21, Records of the Agricultural Workers Health Associations, Records of the Office of Labor, War Food Administration, Agricultural Workers Health Associations, Record Group 224, Region XI, National Archives (hereafter cited as WFA, OL, *Clinic Procedure Manual*); Duffy to Goudy, 22 July 1941, FSA General Correspondence, 1935–1942, Record Group 96, National Archives and Records Administration, Federal Records Center, Region XI, Seattle, Wash.

23. Googe to Duffy, 27 August 1940, FSA General Correspondence, 1935–1942, Record Group 96, National Archives and Records Administration, Federal Records Center, Region XI, Seattle.

24. Henry C. Daniels, interview by author, Takoma Park, Md., 21 December 1984.

25. Thomas Garland Moore Jr., interview by author, Sacramento, Calif., 12 February 1984; Henry C. Daniels, interview by author, Takoma Park, Md., 21 December 1984.

26. USDA, Farm Security Administration, *Annual Report of the Chief Medical Officer for the Fiscal Year, July 1, 1939 to June 30, 1940* (Washington, D.C.: GPO, 1940), 22.

27. This issue is discussed in a series of letters from Region XI, including Goudy to Duffy, 7 July 1941; Duffy to Goudy, 22 July 1941; and Googe to Duffy, 12 December, 1939, FSA General Correspondence, 1935–1942, Record Group 96, National Archives and Records Administration, Federal Records Center, Region XI, Seattle, Wash.

28. An example is *California and Western Medicine,* which was the journal for the state medical societies of California, Nevada, and Arizona. Lorin Kerr, an FSA district medical officer, considered the concept "terrific" when he arrived in the Pacific Northwest in 1944. An office co-worker was not so sure, Kerr recalled, and "pretty nearly collapsed. . . He couldn't get on the phone quick enough, called Washington and said, 'My God, you've got a dummy out here. He wants to pull together the representatives of the seven state medical societies!'" Lorin Kerr, interview by author, Washington, D.C., 11 November 1983. See also Williams to Ringle, 17 November 1941 and Googe to Duffy, 12 December 1939, FSA General Correspondence, 1935–1942, Record Group 96, National Archives and Records Administration, Federal Records Center, Region XI, Seattle, Wash.

29. *California and Western Medicine* 60 (5) 238–40, 1944.

30. Allen Koplin and Pauline Koplin, interview by author, Fort Myers, Fla., 2 February 1984.

31. *Monthly Reports of the Agricultural Workers Health and Medical Association,* August 1943, 4, Records of the FSA, Region IX, Record Group 96, Bancroft Library, University of California, Berkeley (hereafter cited as *Monthly Reports* with month and year).

32. *Public Health Reports* 60 (9): 229–49, 1945.

33. Lorin Kerr, interview by author, Washington, D.C., 11 November 1983, and Harold C. Mayers, interview by author, Washington, D.C., 3 February 1984. During World War II, interest in occupational health rose in direct proportion to the importance of maintaining a healthy work force for the war effort. The work of historians Gerald Markowitz and David Rosner indicate that the FSA's medical and legislative activism was also characteristic of the staff of the Division of Labor Standards of the Department of Labor. Gerald Markowitz and David Rosner, "More than Economism: The Politics of Workers' Health and Safety, 1932–1947," *Milbank Memorial Fund Quarterly* 64 (1): 331–54, 1986; David Rosner and Gerald Markowitz, *Deadly Dust: Silicosis and the Politics of Occupational Disease in Twentieth Century America* (Princeton, N.J.: Princeton University Press, 1991).

34. For discussion of racial stereotyping around venereal diseases, see Allen Brandt, *No Magic Bullet: A Social History of Venereal Disease in the United States Since 1880* (Oxford: Oxford University Press, 1987).

35. William Goldner, *The Costs of Medical Care in the Agricultural Workers*

Health and Medical Association (San Francisco: Agricultural Workers Health and Medical Association, 1941), 29. The sampling method used by Goldner looked only at those persons who sought care in the AWHMA and did not adjust for differences in the number of eligible families between California and Arizona. The differences between the two states in monthly per patient costs were significant. In Arizona this figure ranged from $12.42 to $19.18, while in California the figure was $5.67 to $14.38. Cost per physician visit, which does account for denominator adjustment, was $3.64 in Arizona and $3.48 in California. Physicians' office referrals declined from 47 to 33 percent in Arizona in the second year of operation, while in California they declined from 78 to 67 percent (Appendix III). Document located in the personal papers of Henry C. Daniels, loaned to author.

36. Ralph Gregg, "Report of an Investigation of the Agricultural Workers Health and Medical Association" (Fresno, Calif., 1941). Records of the Agricultural Workers Health Associations, Record Group 224, National Archives.

37. Nonsalaried physicians were typically paid $27.00 for a half-day clinic in which 20 or more patients could be seen, hence the $1.35 estimate. WFA, OL, *Clinic Procedure Manual,* 21.

38. It was nothing short of a scandal during both world wars when a substantial fraction of draftees failed their Selective Service exams for health reasons. The FSA responded to the nation's sense of shame by promoting its medical care programs (particularly those preventive programs involving young adults and teenagers) as a means of preserving national vigor and supplying healthy recruits for the war. By doing this, agency leaders gained a small measure of fiscal and political support. FSA clinics and nursing personnel, for example, conducted examinations for the Civilian Conservation Corps and the National Youth Administration. *Public Health Reports* 59(2):1163–84, 1944.

39. Kerr recalled that during the war he began what he claimed to be the agency's first mobile dental clinic. It was located in Idaho and staffed by a Japanese-American dentist, one of the thousands of Nisei incarcerated in encampments in the interior states following the bombing of Pearl Harbor in 1941. Added Kerr proudly, "We got him commissioned as a captain in the Public Health Service! That was something!" Lorin Kerr, interview by author, Washington, D.C., 11 November 1983.

40. *Monthly Reports,* September 1943, 6.

41. In a 1944 speech, Mott noted that FSA camps had only a 1.5 percent loss of available man-days due to illness-related absenteeism, a figure that compared favorably to general industry figures. Frederick D. Mott, "Speech before the Annual Convention of the American Public Health Association," 3 October 1944. Located in Henry C. Daniels papers, loaned to author.

42. "AWHA Audit Report for the period March 4, 1941 to June 30, 1942," 8–15, Reports and Manuals of Agricultural Workers Health Associations, 1941–1944, Record Group 224, National Archives and Records Administration, Federal Records Center, Seattle.

43. WFA, OL, *Clinic Procedure Manual,* 21.

44. For a discussion of the changing views of medical therapeutics in America,

see Charles E. Rosenberg's "The Therapeutic Revolution: Medicine, Meaning, and Social Change in 19th-Century America," and James Harvey Young, "Patent Medicines and the Self-Help Syndrome," in J. Leavitt and R. Numbers (eds.), *Sickness and Health in America* (Madison: University of Wisconsin Press, 1987), 2d ed.

45. "Minutes of the AWHMA, Portland, Ore. Board of Directors Meeting, 4 December 1941," 2, Records of the Agricultural Workers Health Associations, 1941–1947, Record Group 224, National Archives.

46. Gregg, "Report of an Investigation of the Agricultural Workers Health and Medical Association," 9. Also Wayne Rasmussen, *A History of the Emergency Farm Labor Supply Program, 1935–47,* Agricultural Monograph No. 13 (Washington, D.C.: U.S. Department of Agriculture, Bureau of Agricultural Economics, 1951).

47. Belle Glade distinguished itself during the time because it was the site of a large federal camp program with a marked venereal disease problem. The FSA was so concerned with the issue that it requested assistance from the U.S. Public Health Service. A survey by Solomon Axelrod, late professor emeritus from the University of Michigan School of Public Health, documented that poverty, crowded living conditions, inadequate access to health care, and vice in the form of prostitution and alcohol use were the root causes of the venereal disease epidemic. The tragic history of sexually transmitted disease in Belle Glade has continued to the present day. In the mid-1980s the area came under intense scrutiny again when it was noted that the township's incidence of AIDS rivaled that of metropolitan New York and San Francisco. See Michael R. Grey, "Syphilis and AIDS in Belle Glade, Florida, 1942 and 1992," *Annals of Internal Medicine* 116 (4): 329–34, 1992.

48. Harold C. Mayers, interview by author, Washington, D.C., 3 February 1984.

49. John Duffy, "The American Medical Profession and Public Health: From Support to Ambivalence," *Bulletin of the History of Medicine* 45: 1–22, 1979.

50. *Pacific Coast Journal of Nursing* 37 (11): 658–70, 1941.

51. P. A. Kalisch and B. J. Kalisch, *The Advance of American Nursing* (Philadelphia: Lippincott, 1995), and K. Buhler-Wilkerson, "Bringing Care to the People: Lillian Wald's Legacy to Public Health Nursing," *American Journal of Public Health* 83 (12): 1778–86, 1983.

52. Lorin Kerr, interview by author, Washington, D.C., 11 November 1983. For a general discussion of women in the New Deal, see Roger Biles, *A New Deal for the American People* (De Kalb: Northern Illinois University Press, 1991), 193–206.

53. WFA, OL, *Clinic Procedure Manual*, 2–7.

54. "If a family just moved in, we'd check them out to make sure they had them [immunizations]," recalled Pauline Koplin. "One family we went to, they had 10 or 12 children and we went through the whole batch from the little one to the big one. Allen was clever, I thought, starting with the tiniest one and saying, 'Now if you don't cry nobody will mind,' and then all the rest of them would be brave. And

it worked, it worked all the time." Allen Koplin and Pauline Koplin, interview by author, Fort Myers, Fla., 2 February 1984.

55. Lorin Kerr, interview by author, Washington, D.C., 11 November 1983.

56. Henry C. Daniels, interview by author, Takoma Park, Md., 21 December 1984.

57. Allen Koplin and Pauline Koplin, interview by author, Fort Myers, Fl., 2 February 1984.

58. Henry C. Daniels, interview by author, Takoma Park, Md., 21 December 1984.

59. Lorin Kerr, interview by author, Washington, D.C., 11 November 1983. Allen Koplin and Pauline Koplin, interview by author, Fort Myers, Fla., 2 February 1984.

60. USDA, Farm Security Administration, *Community Nursing Handbook* (Washington, D.C.: GPO, 1941), Appendices C and E.

61. *Pacific Coast Journal of Nursing* 37 (11): 658–70, 1941.

62. Whitlock to McIver, 22 September 1939, FSA General Correspondence, 1935–1942, Region XI, Record Group 96, National Archives and Records Administration, Federal Records Center, Seattle, Wash. See also Googe to Williams, 23 August 1940, same source.

63. Home management supervisors maintained statistical accounts of the productivity of the farm wives in their gardening and canning activities. Gene Leach, *Handbook for Camp Home Management Supervisors Region XI* (Washington, D.C.: USDA, Farm Security Administration, 1963), Record Group 96, Region XI, National Archives.

64. Ibid., 4–8. Many migrant children attended school regularly for the first time in the government camps. Although the FSA was unable to document through statistical data its belief that improved nutrition resulted in improved educational outcomes, modern evidence from the federally funded Head Start program bears this conviction out. Through its relationship with the U.S. Department of Agriculture, home management supervisors obtained surplus food commodities for use in the migrant camp kitchens and the camp's school lunch program. Home management supervisors requested funding to buy families refrigeration units, suggesting that they would lessen the risk of gastrointestinal infections and serve as a valuable educational tool for migrant women. This request would have been impossible to consider prior to the electrification of rural communities through such New Deal programs as the Rural Electrification Administration and Tennessee Valley Authority. Gregg to Magleby, 24 May 1941, FSA General Correspondence, 1935–1942, Record Group 224, Region XI, National Archives and Records Administration, Federal Records Center, Seattle, Wash.

65. Duffy to Alexander, 4 August 1938; Duffy to Alexander, 18 August 1938, and Duffy to Case, March 4, 1938, FSA General Correspondence, 1935–1942, Record Group 224, Region XI, National Archives and Records Administration, Seattle, Wash.

66. USDA, Farm Security Administration, *Community Nursing Handbook,* Exhibit H.

67. Thomas Garland Moore Jr., interview by author, Sacramento, Calif., 12 February 1984. See also Leach, *Handbook for Camp Home Management Supervisors,* and *Monthly Reports,* February 1944, 5.

68. With Sydenstricker, Goldberger completed a classic social epidemiological study of tenant farmers and cotton mill workers using a longitudinal household survey that was an early model for community-based chronic disease epidemiology. Two social histories on pellagra are recommended: Elizabeth Etheridge, *The Butterfly Caste: A Social History of Pellagra* (Westport, Conn.: Greenwood Press, 1972), and Daphne A. Roe, *A Plague of Corn: The History of Pellagra* (Ithaca, N.Y.: Cornell University Press, 1973). Etheridge credits the FSA's emphasis on such practical steps as dietary diversification, home canning, and home gardening, for having a definite impact on the region's unique scourge. This feat was accomplished through the work of county and state health departments, the U.S. Public Health Service, and several private organizations, notably the Red Cross. In addition to the U.S. Public Health Service, several New Deal work relief agencies are credited with ending pellagra in the South, including the FSA, the Civilian Conservation Corps, the U.S. Forest Service, and the Works Progress Administration. For a short biographical article on Goldberger, see Joanne G. Elmore and Alvin R. Feinstein, "Joseph Goldberger: An Unsung Hero of American Clinical Epidemiology," *Annals of Internal Medicine* 121(9): 372–75, 1994.

69. Etheridge, *Butterfly Caste,* 204–5. Brown to Picard, 25 August 1942, Health and Hygiene files, FSA Correspondence, 1935–1942, Region IX, Record Group 96, National Archives. See also Olaf Larson, *Ten Years of Rural Rehabilitation* (Bombay, India: Association of Advertisers and Printers, 1950), 89–90. The data referred to in each source were promulgated by the FSA in promotional brochures and press releases, and were also published in serial articles carried in the official publication of the USPHS, *Public Health Reports.*

70. These topics are covered in great detail in both the *Community Nursing Handbook* and in the monthly statistical summaries provided by field nurses to their supervisors. This effort was reinforced by the home management staff and managers in the farm labor camps and rehabilitation supervisors. The New Deal funded the building of approximately 3 million privies and thousands of sewer systems between 1933 and 1942. This was a part of Roosevelt's jobs creation approach to New Deal relief. Workers supplied by the Civilian Conservation Corps, the Works Progress Administration, and state and local public health departments worked with the FSA in sanitation. There were other health-related activities of the New Deal that involved numerous agencies. The WPA spent nearly $9 billion in direct funds on a massive national construction and employment program. Over its institutional lifetime it constructed or repaired 125,000 buildings and built 78,000 bridges. A total of 572,000 rural roads were built or improved, while nearly 70,000 streets in urban areas were paved, along with 24,000 miles of sidewalks. Over 350 new airports were constructed with WPA funds, and some 8,000 new parks were built. In addition, the WPA provided the money and manpower to build numerous sanitation and water treatment facilities. According to Collins and Tibbetts, the WPA used $40 million of its total $9 billion budget to build hospitals,

sanitaria, and clinics—approximately 400 hospital-related projects in all. The WPA conducted 185,000 health lectures; employed 7,500 nurses; and conducted 6 million eye, ear, tooth, and general examinations. Of course, such a major undertaking brought employment benefits to nurses, physicians, and health educators. In many New Deal social welfare programs, overlapping jurisdiction and conflicting mandates and duties were the rule rather than the exception. Public, private, and local agencies were often jointly involved in economic relief and rehabilitation efforts; in large-scale construction projects; and in educational programs. Selwyn Collins and Clark Tibbetts, *Research Memorandum on Social Aspects of Health in the Depression* (New York: Social Science Research Council, 1937), 116, 127, 167. For a quick overview of the bricks and mortar accomplishments of the WPA, see Biles, *A New Deal*, 104–9.

71. *Journal of the American Medical Association* 111 (9): 763–66, 1938; *Journal of the American Medical Association*, organizational section 114 (1): 45–55, 1940; *Journal of the American Medical Association* organizational section 135 (15): 1013, 1947. From 1941 through the end of the war, commentary on the migrant health programs in most medical journals was spare.

72. Henry Daniels, interview by author, Takoma Park, Md., 21 December 1984. *California and Western Medicine* 49 (3): 226, 1938. This same theme was sounded in an earlier article by Walter Dickie in the same journal, 47 (2): 107, 1937.

73. *California and Western Medicine* 60 (5): 238–40, 1944.

74. Ibid., 51 (4): 272–74, 1939.

75. Texas Farm Laborers Health Association, Board of Directors Minutes, subfile Regular Meeting, [no date], Records of the WFA, Record Group 224, National Archives. See also *California and Western Medicine* 48(6): 397, 1938 and 49 (3): 226, 1938.

76. *California and Western Medicine* 49 (3): 226, 1938.

77. Saum to Peet, 29 December 1939, Records of the WFA, Region XI, Record Group 224, National Archives and Records Administration, Federal Records Center, Seattle, Wash.

78. Googe to Williams, 23 August 1940, FSA General Correspondence Files, 1935–1942, Region XI, Record Group 96, National Archives and Records Administration, Federal Records Center, Seattle, Wash.

79. Coffey to Williams, 13 October 1939, FSA General Correspondence Files, 1935–1942, Region XI, Record Group 96, National Archives and Records Administration, Federal Records Center, Seattle, Wash.

80. Williams to Coffey, FSA General Correspondence Files, 1935–1942, Region XI, Record Group 96, National Archives and Records Administration, Federal Records Center, Seattle, Wash.

81. Mott to Maddox, 19 March 1942. Letter located in Henry C. Daniels papers, loaned to author. Also, Williams to Coffey, FSA General Correspondence Files, 1935–1942, Region XI, Record Group 96, National Archives and Records Administration, Federal Records Center, Seattle, Wash.

82. Sussman to Peet, 13 December 1938, Records of the WFA, Record Group 224, Region XI, National Archives and Records Administration, Federal Records

Center, Seattle, Wash. Also, "Annual Meetings of the Board of Directors, Texas Farm Laborers Health Association," for the following dates: 2 July 1945, 28 April 1946, and 28 March 1943, Records of the Texas Farm Laborers Health Association, Record Group 224, National Archives. Additional physician comment on low fee schedules in *Northwest Medicine* 44 (10): 329, 1945.

83. *California and Western Medicine* 49 (6): 475, 1938.

84. Ibid., 46 (6): 422, 1937.

85. Ibid., 46 (1): 2, 1937. For additional early comments on the FSA programs from the perspective of the California Medical Association, see *California and Western Medicine* 48 (6): 397, 460–61, 1937. For a discussion of the CMA's role in compulsory state insurance plans, see Robert E. Burke, *Olson's New Deal for California* (Berkeley: University of California Press, 1953), and Arthur J. Viseltear, "Compulsory Health Insurance in California, 1934–35," *American Journal of Public Health* 61 (10): 2113–26, 1971.

86. The changing political climate of the FSA from 1938 through the war is covered in greatest detail in Baldwin, *Poverty and Politics,* 295–364. A more current and condensed analysis of the unraveling of the liberal domestic political agenda is contained in Biles, *A New Deal,* 136–53; and Robert S. McElvaine, *The Great Depression: America, 1929–1941* (New York: Times Books, 1984), 306–22.

87. Goldner, *The Costs of Medical Care in the Agricultural Workers Health and Medical Association,"* 4–13.

88. Ibid., 34–35.

89. Ibid., 13–19, 39–77.

90. Ibid., 13–77.

91. The phenomenon of small areas variation has also attracted historians, who have identified it as early as the Progressive Era. Joel D. Howell and C. G. McLaughlin, "Regional Variations in 1917 Health Care Expenditures," *Medical Care* 27 (8): 772–87, 1989. This work, as well as my own on the FSA, suggests that differences in physicians' behavior may be a fundamental characteristic of medical practice in the United States.

92. Gregg, "Report of an Investigation of the Agricultural Workers Health and Medical Association," 1.

93. Ibid., 17–26.

94. Ibid., 12–21.

95. Ibid., 14–15.

96. Ibid., Appendix III.

CHAPTER FOUR: MODELING FOR NATIONAL HEALTH INSURANCE

Epigraph: Carl C. Taylor, T. Wilson Longmore, and Douglas Ensminger, *The Experimental Health Program of the United States Department of Agriculture* (Washington, D.C.: GPO, 1946), 76–77.

1. Much of the text of this chapter follows from Carl C. Taylor, T. Wilson Longmore, and Douglas Ensminger, *The Experimental Health Program of the United States Department of Agriculture,* a study prepared for the Subcommittee on

Wartime Health and Education of the Senate Committee on Education and Labor, 79th Cong., 2d sess. (Washington, D.C.: GPO, 1946), 76–77. Some of the authors of this Senate monograph were employees of the Bureau of Agricultural Economics and prepared the report with the assistance of the FSA for the Senate subcommittee. Except where noted, all statistics are taken from this monograph, cited as Senate Committee, *The Experimental Health Program.*

2. Frederick D. Mott, *Health Provisions in a Social Security Program for Farm People, With Special Reference to the Experience of the Health Programs of the Farm Security Administration* (Washington, D.C.: U.S. Department of Agriculture, Farm Security Administration, 1944), 3, file 31-2, Frederick Dodge Mott Papers, Canadian National Archives, Ottawa.

3. Senate Committee, *The Experimental Health Program,* 2.

4. Mott to Medical Care Staff, 12 December 1942, file 31-9, Frederick Dodge Mott Papers, Canadian National Archives, Ottawa.

5. Mott and many of his colleagues shared the view common at the time that specialization was an appropriate response to the evolving scientific basis of medicine, and therefore it was an essential issue that must be addressed if health services were to be improved, particularly in rural areas. This did not undermine their equally strong commitment to the central role of the general practitioner in any organized system of medical care. Mott, *Health Provisions,* 33; USDA, Interbureau Committee to Coordinate Post-War Programs, *Experimental Rural Health Program, Report of the Interbureau Committee to Coordinate Post-War Programs, March 1942* (Washington, D.C.: GPO, 1942), 1–12.

6. Mott, *Health Provisions,* 15.

7. USDA, Interbureau Committee to Coordinate Post-War Programs, *Report on Activities from July through October 1942,* 8, file 31-1, Frederick Dodge Mott Papers, Canadian National Archives, Ottawa.

8. C. Cochran to W. Cowen (n.d.), in Clay Cochran, *Preliminary Docket on Medical Care and Health Education in the State of New Mexico,* 17. Personal papers of Henry C. Daniels, loaned to author. Sources were particularly rich for Taos, and included materials provided to the author by program participants that were not found in the other archival records. See also T. Wilson Longmore and Theo. L. Vaughan, *Taos County Cooperative Health Association, 1942–43* (Little Rock, Ark.: November 1944), 51, file 32-2, Frederick Dodge Mott Papers, Canadian National Archives, Ottawa. Taos is also discussed in Mott and Roemer, *Rural Health and Medical Care,* 400–405; and in T. Harding Swann, "Better Health for Country Folks: II. In the Mountains of New Mexico," *Survey Graphic* 34: 374–75, September 1945; and Agnes E. Meyer, "Health for Taos," *Washington Post,* April 25, 26, 27, 29, 1946.

9. Cochran, *Preliminary Docket on Medical Care,* 15–17.

10. The FSA, like the New Deal of which it was a part, has been chastised for doing less than it might have for minorities. Pressure to maintain high levels of loan repayment and the racism of some local rehabilitation supervisors resulted in blacks receiving fewer and smaller loans and fewer capital advances than white farmers. Nonetheless, loans to African Americans were granted, and blacks did

benefit from FSA programs. Even critics of the FSA acknowledge that the agency improved the socioeconomic status of some of the nation's poorest African American farmers. Historian Theodore Saloutos, for one, notes that by 1939 the FSA had given out some 50,000 rehabilitation loans to black farmers and resettled 1,000 black families in 31 homestead communities. By standards of the day, states Saloutos, this represented an "extensive program of aid and rehabilitation" for black families. A less quantitative but no less real measure of the FSA's impact on race issues was the degree of enmity the agency and its programs earned from many southern politicians and entrenched agricultural power groups in the South. Theodore Saloutos, *The American Farmer and the New Deal* (Ames: Iowa State University Press, 1982), 154. On the New Deal, the FSA, and race issues, see also Roger Biles, *A New Deal for the American People* (De Kalb: Northern Illinois University Press, 1991), 172–92, and Sidney Baldwin, *Poverty and Politics* (Chapel Hill: University of North Carolina Press, 1968), 282–94.

11. Senate Committee, *Experimental Health Program,* 120.

12. Ibid., 116–47. See also Frederick D. Mott and Milton I. Roemer, *Rural Health and Medical Care* (New York: McGraw-Hill, 1948), 402–3.

13. Williams's letter is cited verbatim in R. Kimmel to All Regional Chairman, 13 January 1942, Health and Welfare files, Records of the Farm Security Administration, Region IX, Bancroft Library, University of California, Berkeley. See also USDA, Interbureau Committee to Coordinate Post-War Programs, *Experimental Rural Health Program,* 5.

14. Senate Committee, *Experimental Health Program,* 12–15, 56, 131–33.

15. Ibid., 29–33, 56, 114.

16. Ibid., 19, 121–30. Other experimental health plans had higher income thresholds than that of the Taos plan. See Longmore and Vaughan, *Taos County Cooperative Health Association,* 51.

17. Mott and Roemer, *Rural Health and Medical Care,* 407; Senate Committee, *Experimental Health Program,* 25, 135.

18. Mott and Roemer, *Rural Health and Medical Care,* 417; Mott, *Health Provisions,* 15; and Senate Committee, *Experimental Health Program,* 11–25.

19. Senate Committee, *Experimental Health Program,* 37–38; Longmore and Vaughan, *Taos County Cooperative Health Association,* 51.

20. Membership data are from Jesse B. Yaukey, *Activities of an Experimental Rural Health Program in Six Counties during its First Fiscal Year, 1942–1943* (Washington, D.C.: GPO, 1943), file 31-2, Frederick Dodge Mott Papers, Canadian National Archives, Ottawa. See also Senate Committee, *Experimental Health Program,* 37–38, 86–115, and Longmore and Vaughan, *Taos County Cooperative Health Association,* 62–65.

21. Senate Committee, *Experimental Health Program,* 71.

22. Ibid., 22–23.

23. Ibid., 31.

24. Longmore and Vaughan, *Taos County Cooperative Health Association,* 51. There is a growing body of historical criticism of the medical, economic, and ethical issues that caused the displacement of midwives from the birthing room. Two

essays in Judith W. Leavitt and Ronald L. Numbers (eds.), *Sickness and Health in America* (Madison: University of Wisconsin Press, 1985) are particularly relevant: Frances E. Kobrin, "The American Midwife Controversy: A Crisis of Professionalization," 197–205, and Daniel M. Fox and Joyce Antler, "The Movement toward a Safe Maternity: Physician Accountability in New York City, 1915–1940," 490–521. This issue also relates to that of medical optimism discussed more fully in Chapter 5.

25. Senate Committee, *Experimental Health Program,* 40–48.

26. Ibid., 31–36. The estimates of medical care needed came from a measure known as the Lee-Jones criteria. Roger Lee and Lewis and Barbara Jones wrote one of the CCMC reports in which they identified the needed level of services, which, not surprisingly, few Americans were receiving. The Lee-Jones criteria provide extremely high estimates of medical need and buttressed the widely held view that Americans were not getting enough medical care, regardless of income. Roger I. Lee, Lewis Webster Jones, and Barbara Jones, *The Fundamentals of Good Medical Care* (Chicago: University of Chicago Press, 1933).

27. *Journal of the Missouri Medical Association* 40 (8): 259–62, 1944; also USDA, Interbureau Committee to Coordinate Post-War Programs, *Experimental Rural Health Program,* 3.

28. Longmore and Vaughan, *Taos County Cooperative Health Association,* 51.

29. Senate Committee, *Experimental Health Program,* 104–5, 137–41.

30. Ibid., 104. The capitated Wheeler County Rural Health Service is discussed in detail on pages 86–115.

31. Writing about the experimental health plans after the program had ended, Mott and Roemer cite the figures in the text. During the earliest days of the programs, the figures for hospitalization were higher. Mott and Roemer, *Rural Health and Medical Care,* 417–19.

32. Senate Committee, *Experimental Health Program,* 104.

33. Ibid., 35–36.

34. The FSA noticed that a similar problem developed when it began the farmers mutual aid corporations. U.S. Resettlement Administration, *Annual Report of the Resettlement Administration* (Washington, D.C: GPO, 1937), 5–13. The association between hospitalization, technology, rising health care costs, as well as the technological imperative characteristic of twentieth-century American medicine has been addressed by numerous scholars. Stevens notes that from the turn of the century American hospitals were enamored of diagnostic and treatment technologies, including surgery, and quickly assumed their current character as (to borrow her phrase) "engineering centers" of American medicine. Rosemary Stevens, *In Sickness and in Wealth* (New York: Basic Books, 1994), 262–67. Economist Eli Ginzberg attributes the expensive and expansive growth of the health care industry in the United States to rising specialization and reimbursement that arose as a function of longer medical (i.e., specialty) training and gains in disposable income as well as a willingness on the part of most Americans to spend for medical care. Eli Ginzberg, *The Limits of Health Care Reform: The Search for Realism* (New York: Basic Books, 1978), 99–112. Pertinent historical

considerations of this issue may be found in Joel D. Howell, " Early Use of X-Ray Machines and Electrocardiographs at the Pennsylvania Hospital," *Journal of the American Medical Association* 255: 2320–23, 1986, and Stanley Joel Reiser, *Medicine and the Reign of Technology* (Cambridge, Mass.: Harvard University Press, 1977).

35. Mott to Maddox, 19 March 1942. Located in the personal papers of Henry C. Daniels, loaned to author.

36. Mott, *Health Provisions,* 15.

37. Senate Committee, *Experimental Health Program,* 75–77, 90, 114–15. In an article published in his state medical journal, medical society president Dr. Carl Vohs noted that 73 percent of all Arkansas peanuts were produced by FSA borrowers and that 85 percent of all loans were repaid by the end of the 1942 calendar year. The average number of physicians' calls ranged from 99 to 165 per thousand per month, and collections averaged just shy of 70 percent. *Journal of the Missouri Medical Association* 40 (8): 259–62, 1944.

38. *Journal of the American Medical Association* 118 (17): 1482, 1942. The term *foreign* is chosen on purpose because there were some forms of group practice based on capitation in urban communities, many of which were tied to community organizations serving ethnic or national groups, such as lodge practice, union-based medical service plans, and consumer plans. For a brief discussion of contract practice, see Paul Starr, *The Social Transformation of American Medicine* (New York: Basic Books, 1982), 206–9, and George Rosen, "Contract or Lodge Practice and Its Influence on Medical Attitudes to Health Insurance," *American Journal of Public Health* 67 (4): 374–78, 1977.

39. Senate Committee, *Experimental Health Program,* 92.

40. Ibid., 76–77.

41. Ibid., 17–19.

42. *Journal of the American Medical Association* 118 (17): 1482, 1942.

43. Ibid., 120 (16): 1315–24, 1942.

44. Lorin Kerr, interview with author, Washington, D.C., 11 November 1983.

45. As cited in *Medical Care* 3 (3): 270, 1943.

46. Senate Committee, *Experimental Health Program,* 77.

47. Mott to Health Services Staff, 8 September 1943, file 31-9, Frederick Dodge Mott Papers, Canadian National Archives, Ottawa.

CHAPTER FIVE: THE FSA GOES TO WAR

Epigraph: Journal of the American Medical Association 114 (2): 159–67, 1940.

1. "It was the more senior of the physicians who carried the load," recalled John Newdorp. "Some counties did not have enough physicians to go into the service [and] there were places where. . . the youngest physician was sixty-five years old!" John Newdorp. interview by author, Alexandria, Va., 3 February 1984. Between 1923 and 1930, twenty-one states (eighteen of which were predominantly rural) lost one-fifth of their physicians. By 1931, a mere 20 percent of medical school graduates practiced in rural areas. Just a decade earlier, this figure had been nearly 50 percent. Frederick D. Mott, *Health Provisions in a Social Security Program*

for Farm People, with Special Reference to the Experience of the Health Programs of the Farm Security Administration (Washington, D.C.: U. S. Department of Agriculture, Farm Security Administration, 1944), 3–4, file 31-2, Frederick Dodge Mott Papers, Canadian National Archives, Ottawa. In 1944, the *Journal of the Missouri Medical Association* commented that one county in the southeastern part of the state lost all but one of its doctors, resulting in a physician-to-population ratio of 1 to 53,000. *Journal of the Missouri Medical Association* 41 (6): 116–18, 1944; see also *Journal of the American Medical Association,* organizational section 131 (6): 538–39, 1945.

2. Sidney Baldwin, *Poverty and Politics: The Rise and Decline of the Farm Security Administration* (Chapel Hill: University of North Carolina Press, 1968), 325–35.

3. Kenneth Davis, *FDR, into the Storm, 1937–1940: A History* (New York: Random House, 1993), 358, 362–63.

4. Baldwin, *Poverty and Politics,* 293–395. More recent historical analysis of agricultural politics as it affected the FSA during this period of institutional decline may be found in David Danbom, *Born in the Country* (Baltimore: Johns Hopkins University Press, 1995), 195–206.

5. Baldwin, *Poverty and Politics,* 286–96, 383–95; Frederick D. Mott and Milton I. Roemer, *Rural Health and Medical Care* (New York: McGraw-Hill, 1948), 406.

6. During the war when the FSA transported and recruited southern blacks to other agricultural regions, southern politicians rammed through a bill forbidding the recruitment of migrant workers from the South. The so-called Pace Amendment generated enormous controversy. Ironically, its negative impact on farm labor recruitment led growers in the western states to join with the FSA's usual allies in a successful effort to pressure Congress to reverse the law. Wayne Rasmussen, *The History of the Farm Labor Programs of the U.S. Department of Agriculture, 1935–1947* (Washington, D.C.: U.S. Department of Agriculture, 1948).

7. Baldwin, *Poverty and Politics,* 356.

8. Walter J. Stein, *California and the Dustbowl Migration* (Westport, Conn.: Greenwood Press, 1973), 211–12.

9. Baldwin, *Poverty and Politics,* 320–22, 345. For the Tolan Committee, see U.S. House, *National Defense Migration: Hearings of the Select Committee to Investigate National Defense Migration,* 77th Cong., 2d sess. (Washington, D.C.: GPO, 1942).

10. War Food Administrator (WFA) Chester Davis came to this new position from his earlier work as AAA Administrator. Baldwin, *Poverty and Politics,* 387–90.

11. Allen Koplin and Pauline Koplin, interview by author, Fort Myers, Fla., 2 February 1984.

12. USDA, Farm Security Administration, *A Handbook on Health for Farm Families,* publication no. 129 (Washington, D.C.: GPO, 1944), 1–6. Document also located in file 31-5, Frederick Dodge Mott Papers, Canadian National Archives, Ottawa.

13. For FSA communities, rejections averaged 23 percent, while for rural communities as a whole 36 percent of draftees were found unfit for military service. U.S. House, *Medical Care Program of the Farm Security Administration: Hearings before the Select Committee of Health Care on Agriculture to Investigate the Activities of the FSA,* 78th Cong., 1st sess. (Washington, D.C.: GPO, 1944). Document also located in file 31-5, Frederick Dodge Mott Papers, Canadian National Archives, Ottawa.

14. Wayne Rasmussen, *A History of the Emergency Farm Labor Supply Program, 1943–1947,* Agricultural Monograph no. 13 (Washington, D.C.: Bureau of Agricultural Economics, U.S. Department of Agriculture, 1951).

15. Bruton to Parran, 7 July 1943, files of the Agricultural Workers Health Associations, Record Group 224, National Archives.

16. *California and Western Medicine* 58 (6): 377, 385, 1943; 59 (5): 287–89, 1943; 60 (1): 32, 1944; and 60 (5): 238–40, 1944. Mott, "Medical Care Program of the Farm Security Administration," file 31-5, and Mott to Health Services Staff, 5 October 1943, file 31-9, Frederick Dodge Mott Papers, Canadian National Archives, Ottawa.

17. *Journal of the American Medical Association* 108 (18): 1524–26, 1937.

18. Ibid., 114 (2): 159–67, 1940.

19. Ibid., 120 (16): 1315–24, 1942.

20. *Journal of the Medical Society of New Jersey* 37 (7): 394–97, 1943. This same point is made in *Journal of the Missouri Medical Association* 40 (8): 259–62, 1943, and *California and Western Medicine* 58 (6): 377, 385, 1943.

21. *Journal of the American Medical Association* 114 (2): 162–64, 1940. On physicians' income see *Journal of the Michigan Medical Society* 46: 190, 1947.

22. *Medical Care* 2 (4): 272–73, 1942.

23. For example, see *Journal of the Medical Association of the State of Alabama* 11(6): 440–41, 1942.

24. *Journal of the Medical Association of the State of Alabama* 11 (12): 440–41, 1942.

25. Physicians objected to a plan which, in the words of the California Medical Association, "forces the physician to contract with the Federal Government for both medical services and hospitalization at fixed fees and the enormous amount of paper work and potential delay in payments." *California and Western Medicine* 59 (4): 206, 226–31, 1943.

26. *California and Western Medicine* 60 (1): 31–32, 1944, and *Journal of the American Medical Association* 129 (17): 1213, 1945. Over a million pregnancies, or nearly one in seven births during the war, received medical attention under provisions of the EMIC. See Nathan Sinai and Odin W. Anderson, *EMIC Emergency Maternity and Infant Care* (Ann Arbor: School of Public Health, University of Michigan, 1948).

27. *Journal of the American Medical Association,* organizational section 116 (25): 2781–85, 1941.

28. USDA, Farm Security Administration, *Annual Report of the Chief Medical Officer for the Fiscal Year July 1, 1940–June 30, 1941* (Washington, D.C.: GPO, 1941), 1. Also Mott, *Health Provisions,* 26–27.

29. *Journal of the American Medical Association,* organizational section 120 (16): 1315–24, 1942.

30. Ibid., 114 (26): 2559, 1940.

31. Ibid., 116 (25): 2784, 1941.

32. Baldwin, *Poverty and Politics,* 391–400.

33. Ibid. Bickering between the British and U.S. governments delayed Baldwin's appointment as the first proconsul to liberated Italy, and instead he accepted a position as an assistant to Sidney Hillman in the Congress of Industrial Organizations (CIO).

34. Mott to Health Services Staff, 25 November 1943, file 31-5, Frederick Dodge Mott Papers, Canadian National Archives, Ottawa.

35. Paul Starr, *The Social Transformation of American Medicine: The Rise of a Sovereign Profession and the Making of a Vast Industry* (New York: Basic Books, 1982).

36. Charles E. Rosenberg, *Explaining Epidemics* (Cambridge: Cambridge University Press, 1992), 258–77.

37. Starr, *Social Transformation,* 235–378; Rosemary Stevens, *American Medicine and the Public Interest* (New Haven, Conn.: Yale University Press, 1971), 267–90; Daniel S. Hirshfield, *The Lost Reform: The Campaign for Compulsory National Health Insurance from 1932 to 1943* (Cambridge, Mass.: Harvard University Press, 1970); and Rickey Hendricks, *A Model for National Health Care: The History of Kaiser-Permanente* (New Brunswick, N. J.: Rutgers University Press, 1993), 78–82. Historical consideration by individuals close to the action include Milton I. Roemer, "I. S. Falk, the Committee on the Costs of Medical Care, and the Drive for National Health Insurance," *American Journal of Public Health* 75 (8): 841–48, 1985, and I. S. Falk, "Proposals for National Health Insurance in the USA: Origins and Evolution, and Some Perceptions for the Future," *Milbank Memorial Fund Quarterly* 55 (2) 161–91, 1977.

38. *Journal of the Arkansas Medical Society* 42 (4): 64–72, 1945.

39. U.S. Bureau of the Census, *Statistical History of the United States, Colonial Times to the Present* (Washington, D.C.: GPO, 1965), 677.

40. *California and Western Medicine* 65 (2): 85, 1946.

41. Other rural health initiatives are summarized in Mott and Roemer, *Rural Health and Medical Care,* 439–69.

42. Helen Johnson, interview by author, Washington, D.C., 29 April 1994. The California Grange and Farm Bureau successfully lobbied the California Medical Association to extend its CPS program to their members during the 1940s. *Journal of the American Medical Association* 130 (10): 649, 1946.

43. *Journal of the American Medical Association* 129 (17): 1187–89, 1945.

44. Ibid. See related comments in *Journal of the American Medical Association* 131 (11): 781–84, 1946; *California and Western Medicine* 50 (2): 143, 1939; and 49 (5): 394, 1938. According to Mott and Roemer, both the Grange and the American Farm Bureau Federation were on record against "any form of socialized medicine which would be administered by any branch of Government." Mott and Roemer, *Rural Health and Medical Care,* 468.

45. *Journal of the American Medical Association* 131 (11): 915–16, 1946. Mott and Roemer, *Rural Health and Medical Care,* 439–48.

46. Statistics culled from the following sources: *Journal of the American Medical Association* 131 (11):781–84,1946; *Medical Care* 4 (2): 17–36, 1947; Mott and Roemer, *Rural Health and Medical Care,* 441–44.

47. Mott to Health Services Staff, 11 October 1945, file 31-9, Frederick Dodge Mott Papers, Canadian National Archives, Ottawa.

48. *California and Western Medicine* 56 (6): 371–72, 1942; 60 (4):192, 1944 and 65 (2): 85, 1946. Beginning in January 1943, the Hospital Savings Association and Hospital Care Association in North Carolina, a Blue Cross–affiliated hospital insurance program, agreed to provide hospital coverage for FSA borrowers at $122 per family. Although North Carolina was the first state to offer a Blue Cross option for FSA borrowers, this arrangement was available to FSA families in a few Oregon and New York counties as well. Approximately 6,000 FSA families enrolled in the North Carolina plan for a total plan membership of 36,000 from 1944 to 1946. A surgical plan was also available to FSA borrowers for an annual fee of $8 per year through the privately sponsored Medical Service Association of Durham, North Carolina. This plan paid 100 percent of surgical fees. By 1946, it had enrolled some 31,000 FSA members. In the fall of 1945, a similar insurance program for physicians' office visits was offered to FSA borrowers for an annual fee of $20, although subscribers to this plan numbered fewer than 6,000 by 1946. Frederick D. Mott, "Prepayment Health Service Plan for FSA Borrowers in N.C.," 10 February 1946, file 31-5, Frederick Dodge Mott Papers, Canadian National Archives, Ottawa. See also *North Carolina Medical Journal* 5 (5): 217–19, 1946, and Mott and Roemer, *Rural Health and Medical Care,* 397–400.

49. *Journal of the American Medical Association* 127 (2): 92–93, 1945.

50. Ibid., 131 (6): 538–39, 1945.

51. Ibid., 131 (11): 915, 1946.

52. *Northwest Medicine* 44 (10): 329, 1945 and 42 (11): 333–35, 1943; *Journal of the American Medical Association* 131 (11): 915–16, 1946.

53. Leslie Falk, interview by author, Shelburne, Vt., 19 February 1994.

54. Mott to Health Services Staff, 11 October 1945, file 31-9, Mott Papers.

CHAPTER SIX: RUNNING AGAINST THE TIDE

Epigraph: Mott to Health Services Staff, 31 May 1946, file 31-9, Frederick Dodge Mott Papers, Canadian National Archives, Ottawa.

1. For a more focused analysis of the national health insurance movement up to the 1943 Wagner-Murray-Dingell bill, see Daniel Hirshfield, *The Lost Reform: The Campaign for Compulsory National Health Insurance from 1932 to 1943* (Cambridge, Mass.: Harvard University Press, 1970). The U.S. Public Health Service's role in the national health insurance debate is briefly summarized by Fitzhugh Mullan, *Plagues and Politics: The Story of the United States Public Health Service* (New York: Basic Books, 1989). The debate over universal health insurance and the state may be found in Ronald L. Numbers, *Almost Persuaded* (Baltimore: Johns Hopkins University Press, 1978), and David J. Rothman, "A Century of Failure:

Health Care Reform in America," *Journal of Health Politics, Policy, and Law* 18 (2): 271–85, 1993. A terse and readable monograph on the subject of national health insurance as part of broader social welfare developments is Edward D. Berkowitz, *America's Welfare State: From Roosevelt to Reagan* (Baltimore: Johns Hopkins University Press, 1991).

2. Some would say that as private health insurance plans became a significant presence in health care, it became increasingly unlikely that a public system could evolve. When Canada developed its public-private system, health insurance companies covered less than 4 percent of the population, and as the public insurance system matured, the private health insurance sector remained vestigial. In the United States insurance companies played an important role, alongside their allies in organized medicine, in defeating the 1947 Truman proposal. Prior to that point, they were not a major influence in the national health insurance debate. After Truman's national health legislation was soundly rejected by Congress, the federal government's obligation to guarantee universal access to medical care simmered as a political issue. Nearly two decades later, a partial victory for reformers came in the form of Medicare and Medicaid. Rosemary Stevens, *American Medicine and the Public Interest* (New Haven, Conn.: Yale University Press, 1976), 272–77.

3. Mott to Health Services Staff, 5 October 1943, file 31-9, Frederick Dodge Mott Papers, Canadian National Archives, Ottawa.

4. Mott to Health Services Staff, 25 November 1943, file 31-9, Frederick Dodge Mott Papers, Canadian National Archives, Ottawa.

5. Leslie Falk, interview by author, Shelburne, Vt., 19 February 1994.

6. Mott to Health Services Staff, 5 October 1943, file 31-9, Mott Papers. Also USDA, Farm Security Administration, *A Handbook on Health for Farm Families,* publication no. 129 (Washington, D.C.: GPO, 1944), 1–6.

7. Frederick D. Mott, "Statement of the Department of Agriculture on the Proposed National Health Program (Senate Bill 1606, the National Health Bill of 1945), 25 April 1945," 8, file 31-20, Frederick Dodge Mott Papers, Canadian National Archives, Ottawa.

8. B. W. Strauss, "The New Deal: When the Motive Wasn't Money," *Washington Monthly,* May 1985, 17–19.

9. John Newdorp, interview by author, Alexandria, Va., 3 February 1984.

10. Lorin Kerr, interview by author, Washington, D.C., 11 November 1983.

11. Samuel Lubell and Walter Everett, "Rehearsal for State Medicine," *Saturday Evening Post,* 17 December 1938.

12. Carl C. Taylor, T. Wilson Longmore, and Douglas Ensminger, *The Experimental Health Program of the United States Department of Agriculture,* a study made for the Subcommittee on Wartime Health and Education of the Senate Committee on Education and Labor, 79th Cong., 2d sess. (Washington, D.C.: GPO, 1946), 1-5. See also John Newdorp, "Planning for Medical Care in the Post-War Period," *Journal of the Medical Association of the State of Alabama* 14 (8): 183–89, 1945.

13. Mott to Health Services Staff, 8 September 1943, file 31-9, Frederick Dodge Mott Papers, Canadian National Archives, Ottawa.

14. John Newdorp, interview by author, Alexandria, Va., 3 February 1984.

15. *Medical Care* 3 (4): 251, 351, 1943.

16. Comments on the FSA were much more spare during the war. I suspect this was due to irregularity in meetings of state and county medical societies and the fact that journals gave a lot of attention to the war-related activities of their members and other war matters. I also observed that the paper used in journals was of lower quality, and the journals were considerably thinner due to wartime rationing. For example, see *Journal Lancet* 58 (7): 228, 318, 1943.

17. Paul Starr, *The Social Transformation of American Medicine: The Rise of a Sovereign Profession and the Making of a Vast Industry* (New York: Basic Books, 1982), 279. For a discussion of FDR's transition from "Dr. New Deal" to "Dr. Win the War," see Doris Kearns Goodwin, *No Ordinary Time: Franklin and Eleanor Roosevelt: The Home Front in World War II* (New York: Simon and Schuster, 1994), 482.

18. The schisms between reformers on the various national health proposals of this period are presented in Hirshfield, *Lost Reform,* 138–65; Starr, *Social Transformation,* 280–89, and 347–51. See also James Rorty, *American Medicine Mobilizes* (New York: Norton, 1939) for a view from the perspective of organized medicine.

19. USDA, Farm Security Administration, *A Handbook on Health for Farm Families,* 1–6.

20. Mott, "Statement of the Department of Agriculture on the Proposed National Health Program," 7.

21. Pepper to Anderson, 17 July 1945, letter included in Senate Committee, *The Experimental Health Program,* iii. Les Falk was a staff member on the Pepper committee. Leslie Falk, interview by the author, Shelburne, Vt., 11 February 1994.

22. Ibid., xix.

23. *Journal Lancet* 66 (9): 260, 1946; *Journal of the American Medical Association,* organizational section 131 (6): 1013, 1948.

24. *Journal of the American Medical Association* 131(6): 538–39, 1946. Reminders that the FSA's chief medical officer was a commissioned officer in the Public Health Service—one of the most suspect federal bureaucracies from the point of view of organized medicine—frequently accompanied any mention of Mott in medical journals at the time.

25. Stevens, *American Medicine and the Public Interest,* 273.

26. A similar strategy was used during the Progressive Era when organized medicine condemned the national health proposal sponsored by the American Association for Labor Legislation as a foreign idea, and ever since this has been a powerful tool that opponents have used to discredit national health insurance and its supporters. Ronald L. Numbers, "The Third Party: Health Insurance in America," in J. W. Leavitt and R. L. Numbers (eds.), *Sickness and Health in America* (Madison: University of Wisconsin Press, 1985), 233–47.

27. California became organized medicine's testing ground for this tactic when Governor Earl Warren proposed the creation of a statewide compulsory insurance

plan. With the aid of a public relations firm and a healthy advertising budget, the California Medical Association reversed initially favorable public opinion and defeated the measure handily. Starr, *Social Transformation,* 275–82.

28. In 1946, the AMA sponsored two rural health conferences (not inviting the FSA to either of them). Describing the FSA plans as "universally a flop," the AMA promoted private insurance plans in rural communities in an obvious attempt to blunt any effort by the federal government to reassert itself in this area. The goal of the meetings was to improve rural medical services by relying solely on private insurance and physician-sponsored plans. *Journal Lancet* 66 (9): 260, 1946; *Journal of the American Medical Association* 133 (11): 778–87, 1947, and 133 (12): 860–67, 1947.

29. Vicente Navarro, *Medicine Under Capitalism* (New York: Prodist, 1977); E. Richard Brown, *Rockefeller Medicine Men: Medicine and Capitalism in America* (Berkeley: University of California Press, 1979).

30. Starr, *Social Transformation,* 266–89.

31. R. G. Leland, "Rural Medicine," in *Proceedings of the Conference Held at Cooperstown, New York, October 7–8, 1938* (Springfield, Ill.: Charles C. Thomas, 1939), 236.

32. *Northwest Medicine* 42 (11): 333–35, 1943.

33. *Journal of the American Medical Association* 133 (11): 778–87, 1947.

34. Stevens, *American Medicine and the Public Interest,* 267–89, and Starr, *Social Transformation,* 235–43.

35. If the debate over national health insurance continues to be framed around these terms, then the conflict the issue raises will remain a source of frustration to supporters of national health insurance, and a saving grace for those who oppose it. On Americans' egalitarianism versus defense of individual liberties, see Alexis de Tocqueville, *Democracy in America* (New York: New American Library, 1956).

36. *California and Western Medicine* 67 (4), 269–72, 1947. Allen Koplin and Pauline Koplin, interview by author, Fort Myers, Fla., 2 February 1984. See also, Milton I. Roemer, "Joseph W. Mountin, Architect of Modern Public Health," *Public Health Reports* 108 (6), 727–35, 1993. Truman grew increasingly sensitive to the issue and under pressure from conservatives in both parties, instituted a loyalty program in March 1947. On McCarthyism and Truman, see David McCullough, *Truman* (New York: Simon and Schuster, 1992), 764–70.

37. *Journal of the Arkansas Medical Society* 44 (3): 79, 1947.

38. The poster also identified by name those individuals and public and private organizations that the AMA accused of supporting national health insurance. Both Mott and Roemer were listed on the poster mentioned in the text. Poster reproduction courtesy of Physicians Forum. It is clear from the oral histories that the intolerant attitude in Washington and nationally had much to do with their decisions.

39. *Journal of the American Medical Association* 132 (10): 584–89, 1946.

40. *California and Western Medicine* 67 (4): 269–72, 1947.

41. W. Andrew Achebaum, *Social Security: Visions and Revisions* (Cambridge: Cambridge University Press, 1986), 161–78; Stevens, *In Sickness and in Wealth* (New York: Basic Books, 1994), 84–89.

42. Although Dr. Roemer did not mention it himself, several of his former colleagues interviewed for this project made the point that he was forced out of the Public Health Service because of his progressive political views. During some of the worst years of the McCarthy period, Roemer worked in Saskatchewan. Leslie Falk, interview by author, Shelburne, Vt., 19 February 1994. Allen Koplin and Pauline Koplin, interview by author, Fort Myers, Fla., 2 February 1984.

43. Mott to Douglas, 16 March 1946, file 45-8, Frederick Dodge Mott Papers, Canadian National Archives, Ottawa.

44. Mott to Health Services Staff, 31 May 1946, file 31-9, Frederick Dodge Mott Papers.

45. Sidney Baldwin, *Poverty and Politics: The Rise and Decline of the Farm Security Administration* (Chapel Hill: University of North Carolina Press, 1968), 395–402.

46. *The Farmers Home Administration Act of 1946*, 79th Cong., 2d sess., H.R. 5991.

47. R. Lubove, "The New Deal and National Health," *Current History* 45 (August 1963): 77–86, 117; McCullough, *Truman,* 337–52.

48. Yaukey to Mott, 22 December 1946, file 46-19, Frederick Dodge Mott Papers, Canadian National Archives, Ottawa.

49. Mott to Yaukey, 27 March 1947, file 46-19, Frederick Dodge Mott Papers, Canadian National Archives, Ottawa.

50. *Journal of the Ohio State Medical Society* 43 (5): 528–34, 1947; see also *Journal Lancet* 67 (9): 337, 1947.

51. Indeed, the only actual congressional hearings held on this volatile issue occurred in 1939 with the first Wagner bill, and in 1946 with the 1945 Wagner-Murray-Dingell bill. In addition to previously cited works, see Agnes W. Brewster, *Health Insurance and Related Proposals for Financing Personal Health Services: A Digest of Major Legislation and Proposals for Federal Action, 1935–1957* (U.S. Department of Health, Education and Welfare, Social Security Administration, 1958).

CHAPTER SEVEN: CONCLUSION

Epigraph: Alexander to Mott, 6 June 1945, file 48-26, Frederick Dodge Mott Papers, Canadian National Archives, Ottawa.

1. According to Theodore Saloutos, even "with all its limitations and frustrations, [the New Deal] by making operational ideas and plans that had been long on the minds of agricultural researchers and thinkers, constitute[s] the greatest innovative epoch in the history of American agriculture." Saloutos, "New Deal Agricultural Policy: An Evaluation," *Journal of American History* 61(2): 416, 1974.

2. Edward D. Berkowitz, *America's Welfare State: From Roosevelt to Reagan* (Baltimore: Johns Hopkins University Press, 1991), 153–86.

3. Statisticians George St. John Perrott and Joseph Mountin, among the most senior individuals in the Public Health Service who supported national health insurance, paid a price for their views. A number of individuals with direct or indirect ties to the FSA became involved in a variety of positions that built on their FSA experiences. For example, in the postwar period, E. Richard Weinerman and Milton Roemer became involved in international refugee efforts through the United Nations. Others, such as Allen Koplin and George Silver, took refuge in academic positions in schools of medicine and public health. As noted in the text, Fred Mott left the FSA for Canada, later to be followed by Milton Roemer, and Henry Sigerist left the United States for the more tolerant Switzerland.

EPILOGUE

Epigraph: Thomas Garland Moore Jr., telephone interview by author, Sacramento, Calif., 24 November 1979.

1. Sidney Baldwin, *Poverty and Politics: the Rise and Decline of the Farm Security Administration* (Chapel Hill: University of North Carolina Press, 1968), 4. In the 1970s there was a reinvigorated movement on behalf of the small farmer in the South. The Southern Cooperative Development Fund and other agricultural organizations resurrected the old FSA cooperative approach, but like their predecessor, did not reverse the trend. John T. Kirby, *Rural Worlds Lost: The American South, 1920–1960* (Baton Rouge: Louisiana State University Press, 1987), 349–51.

2. John Newdorp, interview by the author, Alexandria, Va., 3 February 1984. Arthur J. Viseltear, "Emergence of the Medical Care Section of the American Public Health Association, 1926–1948," in *The History of Medical Care in the United States* (Washington D.C.: American Public Health Association, 1972).

3. Jacalyn Duffin, "The Guru and the Godfather: Henry Sigerist, Hugh MacLean, and the Politics of Health Care Reform in 1940s Canada," *Canadian Bulletin of the History of Medicine* 9: 191–218, 1992; Milton I. Roemer, "Prudence in International Comparisons: Insights for the United States from the Canadian Health Insurance Experience," *International Journal of Health Services* 21 (4): 681–84, 1991.

4. The decision to recruit Mott was either the suggestion of Sigerist or Thomas Parran. The Mott family recalls that it was Parran who recommended the FSA chief medical officer to Douglas when the newly elected premier sought Parran's advice on how to put the program together. Andrew Mott to author, 21 February 1996, personal communication. In 1961, Mott was asked to mediate the bitter 1961 physicians' strike in Saskatchewan, but he declined. Mott taught and consulted on the subject of medical care at the University of Toronto until his retirement in 1972.

5. Duffin, "The Guru and the Godfather," 191–218.

6. Eugene Vayda, Robert G. Evans, and William R. Mindell, "Universal Health Insurance in Canada," *Journal of Community Health* 4(3): 217–31, 1979; Harlan D. Dickinson, "The Struggle for State Health Insurance: Reconsidering the Role of Saskatchewan Farmers," *Studies in Political Economy* 41: 133–56, 1993; and Gordon H. Hatcher, "Goals of Public Policy in Canadian Health Insurance Administration," in *Universal Free Health Care In Canada, 1947–77* (Washington, D.C.: U.S. Department of Health and Human Services, 1981), 1–21.

7. Frederick D. Mott, "Recent Developments in the Provision of Medical Services in Saskatchewan," *Canadian Medical Association Journal* 58 (2): 195–200, 1948. Once more, the rest of the nation followed Saskatchewan's lead so that by 1968, all Canadians had hospital and medical care coverage through a federal-provincial insurance partnership.

8. For a history of the UMWA Welfare and Retirement Fund, see Janet E. Ploss, "A History of the Medical Care Program of the United Mine Workers of America Welfare and Retirement Fund" (master's thesis, Johns Hopkins School of Hygiene and Public Health, 1980). See also Daniel Fox and Judith Stone, "Black Lung: Miners' Militancy and Medical Uncertainty, 1968–1972," *Bulletin of the History of Medicine* 54 (spring): 54–64, 1980.

9. John Newdorp, interview by the author, Alexandria, Va., 3 February 1984. Newdorp was Mott's deputy in the Miner's Memorial Hospital Association.

10. Once the Farmers Home Administration was created, a few survivors of the FSA Health Services staff were transferred to the new program. Dr. Henry Makover succeeded Fred Mott as chief medical officer. Makover in turn was replaced by Dr. Mark Ziegler as FHA chief medical officer. Those remaining in the Department of Agriculture hoped that they could maintain the gains made by the FSA plans by transferring them to private, state, or local agencies, but this hope proved illusory. Mark Ziegler, E. Richard Weinerman, and Milton I. Roemer, "Rural Prepayment Medical Care Plans and Public Health Agencies," *American Journal of Public Health* 37 (12): 1578–85, 1947.

11. Frazier credits Kerr and others for their dedication to the issue of black lung and for "sparking" the creation of the Black Lung Association, perhaps the first industrial disease advocacy group of its kind. Claude Frazier, *Miners and Medicine: West Virginia Memories* (Norman: University of Oklahoma Press, 1992), 108–16. See also Leslie Falk, "Group Health Plans in Coal-Mining Communities," *Journal of Health and Human Behavior* 4: 4–13, 1963. Kerr stayed with the UWMA until his death in 1992. In his honor, the Occupational Safety and Health section of the American Public Health Association established an annual award for young health and safety activists. James L. Weeks, "Lorin Kerr, MD, MSPH, 1920–1991," *American Journal of Industrial Medicine* 21: 609–11, 1992.

12. Lorin Kerr, interview by author, Washington, D.C., 11 November 1983.

13. Under Mott's leadership, the organization eventually bought the Detroit Metropolitan Hospital and built two community clinics to provide services to union members and their families. Mott stayed in Detroit until 1964. In 1964, he left Detroit to become a consultant for the New York Academy of Medicine until 1966 when he accepted a professorship at the University of Toronto. Andrew Mott to author, personal communication.

14. Theodore R. Marmor, *The Politics of Medicare* (Chicago: Aldine Publishing, 1970), 5–28.

15. Alice Sardell, "Neighborhood Health Centers and Community-Based Care: Federal Policy from 1965 to 1982," *Journal of Public Health Policy,* December: 484–503, 1983.

16. Ibid. Between 1964 and 1982, the OEO and its immediate successor, the

Community Services Administration, funded some 900 health clinics. Most were in urban centers, although some were in rural communities.

17. The first director of the OEO neighborhood health center program, Lisbeth Bamberger Schorr, worked briefly as an assistant to Fred Mott during his tenure with the UAW programs in Detroit. Schorr does not recall Mott mentioning his FSA experiences, but she does remember that Nelson Cruikshank, who had at one time supervised the FSA's farm labor programs from the Washington office, spoke about the FSA programs in his role as director of Social Security for the AFL-CIO in the 1960s. Lisbeth Bamberger Schorr to author, personal communication. See "White House Conference on Health, November 3–4, 1965," file 30-16, Frederick Dodge Mott Papers, Canadian National Archives, Ottawa; and Frederick D. Mott, "Kenneth Pohlman Memorial Dinner Speech to the Medical Care Section of the American Public Health Association," 2 November 1977, file 46-6, Frederick Dodge Mott Papers, Canadian National Archives, Ottawa.

18. Richard Couto, *Ain't Gonna Let Nobody Turn Me Around: The Pursuit of Racial Justice in the Rural South* (Philadelphia: Temple University Press, 1991), 392.

19. Ibid., 264–85, and Melissa Coolidge Smith, "Sick for Justice" (A.B. thesis, Harvard University, 1980).

Index

The Library of Congress has cataloged the hardcover edition of this book as follows:

Grey, Michael R.
    New deal medicine : the rural health programs of the Farm Security
Administration / Michael R. Grey.
        p.   cm.
    Includes bibliographical references and index.
    ISBN 0-8018-5939-5 (alk. paper)
    1. Rural health services—United States—History.   2. United
States. Farm Security Administration. Health Services Division—
History.   I. Title.
RA771.5.G74   1999
362.1´04257´0973—dc21   98-18240

ISBN 0-8018-6917-X (pbk.)